T0301352

Personal Sustainability Practices

NEW HORIZONS IN SUSTAINABILITY AND BUSINESS

Books in the New Horizons in Sustainability and Business series make a significant contribution to the study of business, sustainability and the natural environment. As this field has expanded dramatically in recent years, the series will provide an invaluable forum for the publication of high-quality works of scholarship and show the diversity of research on organization and the environment around the world. Global and pluralistic in its approach, this series includes some of the best theoretical and analytical work with contributions to fundamental principles, rigorous evaluations of existing concepts and competing theories, stimulating debate and future visions.

Titles in the series include:

Pioneering Family Firms' Sustainable Development Strategies
Edited by Pramodita Sharma and Sanjay Sharma

Personal Sustainability Practices
Faculty Approaches to Walking the Sustainability Talk and Living the UN SDGs
Edited by Mark Starik and Patricia Kanashiro

Personal Sustainability Practices

Faculty Approaches to Walking the
Sustainability Talk and Living the UN SDGs

Edited by

Mark Starik

*Senior Lecturer, Sustainability Management (Program),
University of Wisconsin System, USA*

Patricia Kanashiro

*Associate Professor, Sellinger School of Business and
Management, Loyola University Maryland, USA*

NEW HORIZONS IN SUSTAINABILITY AND BUSINESS

Edward Elgar
PUBLISHING

Cheltenham, UK • Northampton, MA, USA

Cover image: *Disposable Empires*, by Billy Friebele (one of the faculty contributors of this volume), is an exhibition that explores the ideology of consumption. Styrofoam packaging is used in an interactive sculpture where sensors, motors and lights reactivate take-out containers. Animating this harmful material depicts a dystopian future where Styrofoam gains consciousness and outlives us all. The animated version can be seen here: http://www.billyfriebele.com/DisposableEmpires.html

Published by
Edward Elgar Publishing Limited
The Lypiatts
15 Lansdown Road
Cheltenham
Glos GL50 2JA
UK

Edward Elgar Publishing, Inc.
William Pratt House
9 Dewey Court
Northampton
Massachusetts 01060
USA

A catalogue record for this book
is available from the British Library

Library of Congress Control Number: 2021943524

This book is available electronically in the **Elgar**online
Business subject collection
http://dx.doi.org/10.4337/9781800375130

ISBN 978 1 80037 512 3 (cased)
ISBN 978 1 80037 513 0 (eBook)

Printed and bound by CPI Group (UK) Ltd, Croydon, CR0 4YY

To all of the sustainability academics, practitioners, students, and other stakeholders, both whom I have met and whom I haven't met, who are trying to improve the quality of life for current and future human generations and for the rest of the planet—thank you so much! M.S.

To my mom Julia Tinen Kanashiro and in memory of my grandmother Teresa Tinen, role models of resilience and sustainable living. P.K.

Contents

Figures

Tables

About the editors

Mark Starik (Ph.D., University of Georgia) is Senior Lecturer for the University of Wisconsin System, teaching Sustainability Research and Corporate Social Responsibility courses, and is Contributing Faculty member of the Walden University School of Public Policy and Administration doctoral program. He has published dozens of academic and practitioner articles, books, cases, chapters, editorials, and other works on the topics of sustainability, corporate social responsibility, and climate action, and was a founding member of the Academy of Management Organizations and the Natural Environment (ONE) Division. Mark has served as Faculty Advisor of the sustainability-oriented student organization Net Impact at several universities and also as a director of sustainability-related institutes, centers, and programs at those same institutions. In addition to academia, he has worked in a half-dozen U.S. business, government, and non-profit organizations in mid-management and energy researcher capacities, holds several sustainability certifications, has been an active volunteer in several sustainability-oriented non-profit organizations, and has provided financial support to numerous national and international social and animal welfare causes.

Patricia Kanashiro (Ph.D., The George Washington University) is Associate Professor at Loyola University Maryland. Her research interests and publications focus on corporate governance, sustainability strategy, and social and environmental challenges. She serves on the editorial board of the journal *Organization & Environment* and received the Outstanding Service Award from the Academy of Management—Organizations and the Natural Environment (ONE) division. Patricia was awarded a two-year Fulbright scholarship to pursue her master's in international development at the University of Pittsburgh. Prior to joining academia, she worked for more than fifteen years in the private and non-profit sectors, including ABN AMRO Bank, Institutional Shareholder Services (ISS), United Nations–International Labour Organization, and Ethos Institute for Business and Social Responsibility. Patricia is a co-founder of Batalá Washington DC, a non-profit organization aimed to empower women through drumming. She is a sansei (third generation from Japan), born in Sao Paulo, Brazil and lives in Bethesda, MD with her husband and two boys.

About the contributors

Joel Bothello (Ph.D., ESSEC Business School) is Associate Professor in Strategy and Sustainability at the John Molson School of Business, Concordia University and holds the position as the Concordia University Research Chair in Resilience and Institutions. His research interests revolve around how organizations and individuals respond to institutional and market demands. His main project focuses on the townships of Cape Town, South Africa, where he studies informal economy entrepreneurship. He is a co-founder of OS4Future.

Kevin D. Carlson (Ph.D, University of Iowa) is Professor of Management and Associate Dean for Research and Faculty Affairs in the Pamplin College of Business at Virginia Tech. His current interests are workforce effectiveness and the role of problem solving, data analytics, and improvement management in driving systematic improvement in ecosystems and the human condition.

Gary Chapman (Ph.D., Queen's University Belfast) is Senior Lecturer in Business and Management at Leicester Castle Business School, De Montfort University, UK. Gary's research focuses on firm innovation and strategy. He has published in leading journals such as *Research Policy* and *Technovation*.

Gary Cocke (M.S., Texas Christian University) serves as Director of Sustainability and Energy Conservation, leading the Office of Sustainability at UT Dallas. He holds bachelor's and master's degrees in Biology with a concentration on ecology and conservation, and brings experience in municipal and non-profit leadership to his work in higher education sustainability. The belief that everyone has a role to play in sustainability guides his work to integrate sustainability across the campus through student engagement, academics, operations, and administration.

Giuseppe Delmestri (Ph.D., Mannheim University) is Professor of Organization and Management Theory and Head of the Institute of Change Management and Management Development at WU Vienna University of Economics and Business, Austria, where he is also member of the STaR Center for Sustainability, Transformation and Responsibility. He is a co-founder of OS4Future, board member at EGOS and ASSIOA (the Italian association for Organization Studies) and responsible for management topics in the Arbeitsgruppe ProDok Faculty of VHB.

Rick Dickinson (M.B.A., Presidion Graduate School) is a strategic thinker with a long career in sustainable product development, marketing, and communications. Currently, he is a business advisor at the Center for Inclusive Entrepreneurship, helping build business communities with a focus on disadvantaged and low-income populations. He supports pre-venture and early venture businesspeople in the North Olympic Peninsula of Washington State.

Melissa Edwards (Ph.D., University of Technology, Sydney) is Director of the Executive MBA program at the University of Technology Sydney, Business School. She teaches sustainability in the business undergraduate and postgraduate programs and has held various National and University grants for embedding sustainability in curricula. She has widely published on corporate sustainability, social impact, and sustainability in management education.

Helen Etchanchu (Ph.D., ESSEC Business School) is Associate Professor at Montpellier Business School, where she holds the Chair on Communication and Organizing for Sustainability Transformation and leads the pedagogy and research group of the sustainability lab. Her research explores how we may organize for and make sense of grand challenges such as climate change, inequality, or health. She is a co-founder of OS4Future.

Regina Frank (Ph.D., Universität Bremen) is Senior Lecturer in Entrepreneurship and Innovation at Leicester Castle Business School, De Montfort University, UK. Regina's research interests include social innovation and entrepreneurship for poverty alleviation, for and with marginalized communities and in emerging markets, as well as education for sustainable development.

Billy Friebele (M.F.A., Maryland Institute College of Art) is a multimedia artist and Associate Professor of Art at Loyola University Maryland. Working with hybrid combinations of physical forms and digital processes, he explores the balance between technology, humans, and nature. Friebele has exhibited at the Baltimore Museum of Art, the Orlando Museum of Art, the Art Museum of the Americas, and the Kreeger Museum of Art among other venues nationally and internationally.

Joanna Gentsch (Ph.D., University of Texas at Dallas) is the Director of Community Engaged Learning for the Office of Undergraduate Education and a faculty member in the School of Behavioral and Brain Sciences at the University of Texas at Dallas. As a developmental psychologist, she has witnessed first hand the reciprocal benefits of community-engaged learning for students, faculty, staff, and community partners. She enjoys teaching undergraduates and focuses her efforts on developing innovative co-curricular programming.

John H. Grant (D.B.A., Harvard Business School) is a Sustainability Advisor, based in Colorado, with particular interests in organizational performance measures, including interactions with the biosphere and political systems. He has served as Visiting Fellow in the School of Global Environmental Sustainability at Colorado State University (2015–17) and previously as the R. Kirby Professor of Strategic Management at the University of Pittsburgh.

Stefanie Habersang (Ph.D., Leuphana University Lüneburg) is a post-doctoral researcher at the Institute of Management and Organization at the Leuphana University of Lüneburg, Germany. Her research interests revolve around female entrepreneurship and gender stereotypes, sustainable digital transformation and novel qualitative approaches in organization and management research. She is a co-founder of OS4Future.

William E. Hefley (Ph.D., Carnegie Mellon University) is Clinical Professor at the University of Texas at Dallas, USA. His research interests include corporate social responsibility, and he was a member of the US Technical Advisory Group throughout the development of ISO 26000. He is a graduate of Carnegie Mellon (Ph.D., Organization Science & Information Technology and M.S., Engineering & Public Policy), University of Southern California (MSSM), San José State University (B.A.), and Excelsior College (B.S.). He holds multiple professional certifications, including Certified Management and Business Educator (C.M.B.E.).

Mark Heuer (Ph.D., George Washington University) is Associate Professor of Management and Marketing at Susquehanna University. He earned an M.B.A. from the University of Maryland and attended the Lutheran School of Theology at Chicago and Yale Divinity School. Dr. Heuer has over 30 years of professional experience in the corporate, government, and non-profit sectors.

Gabriela Gutierrez Huerter O (Ph.D., University of Nottingham) is Assistant Professor in International Management at King's Business School, King's College London. Her work is positioned at the intersection of international business, corporate social responsibility, and comparative institutional analysis. She is a member of OS4Future.

George Ionescu (Ph.D., Academia de Studii Economice din Bucuresti) is a former stockbroker at the Bucharest Stock Exchange and Professor of Finance at the Romanian–American University in Bucharest, Romania. George has extensive business experience related to international financial markets, investment management, venture capital, and sustainable development. He was co-opted into the Drafting Committee for National Sustainable Development Strategy Romania 2013–2020–2030 and is the author of more than fifty articles published in scientific journals or in proceedings of international conferences.

Jimmy Y. Jia (M.B.A., University of Oxford) has a multidisciplinary background, as an entrepreneur, professor, author, and speaker. He is Visiting Scholar at George Washington University and an adjunct faculty member at Presidio Graduate School and Saybrook University, where he teaches strategy, climate change, and sustainable energy systems courses. He is on numerous boards, including of the Center for Sustainable Energy and of the CleanTech Alliance. He holds five patents, has authored numerous publications, is a frequent speaker on topics of energy futurism and leadership, and is a D.Phil. Candidate at University of Oxford studying climate finance.

Te Klangboonkrong (Ph.D., Cranfield University) is Senior Lecturer in Business and Management at Leicester Castle Business School, De Montfort University, UK. Te's current research involves entrepreneurship by disadvantaged social groups. Te holds an M.B.A. from the Asian Institute of Technology and served as a Client Consulting Analyst at Nielsen in Thailand.

Ralph Meima (Ph.D., Lund University) directs solar project development for Green Lantern Group and his career spans the fields of software development, IT marketing, and consulting. Ralph was the founding director of the Marlboro M.B.A. in Managing for Sustainability, serves on his town's energy committee, advocating for renewable energy and climate action, and has an M.B.A. in Engineering degree. Ralph has lived in six countries and speaks several languages. Since 2003 he has lived in Brattleboro, Vermont with his family and has published a speculative fiction trilogy and several works of non-fiction.

Van V. Miller (Ph.D., University of New Mexico) is Professor Emeritus of Central Michigan University and served as a Fulbright Research Scholar to Central America in 1988. He received two U.S. Department of Education grants to develop sustainability programs at Central Michigan University during the period 2008–2013 and recently finished a NASA grant examining the influence of human decisions upon wildland fire outcomes.

Shelley F. Mitchell (Ph.D., University of New Hampshire) is Professor of Management & Sustainability and Research Fellow at HULT International Business School, USA. At the University of New Hampshire, she taught for eight years at the Paul College of Business and Economics. Her teaching and research span socially responsible business, strategic management, social and environmental entrepreneurship, SMEs and ecological sustainability, sustainability mindset, and sustainability in management education, for which she has received numerous academic awards.

Dave Nelson (Ph.D., University of Toledo) has worked in the fields of

Environmental Science and Health & Safety for 25 years across the U.S., having developed and managed multiple testing laboratories. He has graduate degrees in Science and Business and a Ph.D. in Operations Management, with research areas of Alternative Energy, Supply Chain Sustainability, Corporate Social Responsibility, Circular Economy, Climate Change Adaptation and impact of Coronavirus.

Louise Obara (Ph.D., Cardiff University) is Senior Lecturer in Business and Management at Leicester Castle Business School, De Montfort University, UK. Louise teaches and conducts research in the fields of sustainability, corporate social responsibility, and business and human rights. Previously, Dr. Obara worked as Research Associate at the BRASS Research Centre (for Business Relationships, Accountability, Sustainability and Society) and as a social worker in the public sector.

Bruce Paton (Ph.D., University of California, Santa Cruz) earned an M.B.A. from Stanford University and is currently Professor of Management and Innovation at Menlo College. He helped launch the Sustainable Business program at San Francisco State University, has taught sustainable business courses for nearly twenty years, and has served as Dean for two business schools.

Gordon Rands (Ph.D., University of Minnesota) is Professor of Management at Western Illinois University, co-chair of its campus sustainability committee, and teaches classes in business ethics and social responsibility, managing organizations for environmental sustainability, and other management topics. Gordon is a co-founder and past chairperson of the Organizations and the Natural Environment (ONE) division of the Academy of Management and is a past president of the International Association for Business and Society (IABS).

Carolyn Reichert (Ph.D., Pennsylvania State University) is a Clinical Associate Professor at UT Dallas. She has served on the Sustainability Committee, and her research interests include sustainability, corporate governance, and compensation. She is also a graduate of The Ohio State University (B.S. in Finance and B.A. in Mathematics).

Bernadette Roche (Ph.D., University of North Carolina Chapel Hill) is an Associate Professor of Biology at Loyola University Maryland, where she is also Director of the Environmental and Sustainability Studies Minor. She is studying ecological genetics of a small mustard, *Arabidopsis lyrata*, the lyre-leaved rock cress. This close relative of the model organism, *A. thaliana*, grows throughout the eastern part of North America, in a variety of habitats. Her lab is interested in the extent to which several target populations have become genetically differentiated, and whether the genetic differences are due to local adaptation, driven by natural selection. Dr. Roche obtained her B.A. in Biology at the University of Virginia.

Elke Schüßler (Dr.rer. pol., Freie Universität Berlin) is Professor of Business Administration and Head of the Institute of Organization Science at Johannes Kepler University Linz, Austria. She is a member of OS4Future and currently acts as Vice President of the European Group for Organizational Studies.

Robert Sroufe (Ph.D., Michigan State University) is the Murrin Chair of Global Competitiveness, and Professor of Sustainability at Duquesne University's Palumbo Donahue School of Business. He develops research and project-based pedagogy within the #1 U.S.-ranked M.B.A. Sustainable Business Practices program. Winner of numerous teaching awards, he has published international and national refereed journal articles and multiple books on sustainable business practices.

Thomas E. Stone (Ph.D., University of Maine) is Associate Professor of Physics and the Sustainability Director at Husson University in Bangor, Maine. His research interests broadly include sustainability pedagogy, campus sustainability, and local energy issues. He received his commission as a Naval Officer after graduating with a B.S. in Physics from the United States Naval Academy and completed an M.S. in Physics at the University of Wisconsin–Madison.

Wendy Stubbs (Ph.D., Monash University) is Associate Professor in the Human Geography Discipline in the School of Social Sciences. She has taught Corporate Sustainability at Monash University for over fifteen years in the Master of Environment & Sustainability, the Master of Business Administration (M.B.A.), and the Master of Business. Her research seeks to understand how business can more holistically address its environmental, social, and economic responsibilities.

Amy K. Townsend (Ph.D., Antioch University New England) has authored several books on corporate environmental sustainability, business ecology, and biofuels, and has taught courses at the George Washington University and James Madison University. For more than twenty years, she ran Sustainable Development International Corporation, her own consulting firm, and worked with start-up companies in the areas of extreme energy efficiency and biofuels.

Madhavi Venkatesan (Ph.D., Vanderbilt University) is a faculty member in the Department of Economics at Northeastern University (Boston, Massachusetts). She is also Founder and Executive Director of Sustainable Practices, a Cape Cod-based environmental advocacy non-profit, and moreover holds master's degrees in sustainability from both Harvard University and Vermont Law School. She has published three economics textbooks under the series *A Framework for Sustainable Practices* and her most recent book, *SDG8 – Sustainable Economic Growth and Decent Work for All*, was published in 2019.

Acknowledgements

Alan Sturmer and his colleagues at Edward Elgar Publishing

Our external reviewers

Colleagues in the Academy of Management – Organizations and the Natural Environment (ONE) and Social Issues in Management (SIM) divisions; the International Association for Business & Society (IABS); and the Group of Research on Organizations and the Natural Environment (GRONEN)

Loyola University Maryland

University of Wisconsin System

1. Introduction to *Personal Sustainability Practices*

Mark Starik and Patricia Kanashiro[1]

Welcome to our volume *Personal Sustainability Practices: Faculty Approaches to Walking the Sustainability Talk and Living the UN SDGs*! We are excited about this topic because it is a subject with which we co-editors as well as our authors and readers, has likely wrestled more than once in our respective careers as sustainability educators: *"Are we walking our sustainability talk? Are enough of those of us who are faculty members practicing what we preach as frequently, substantively, and effectively as we should on the topic of sustainability?"* Additionally, and importantly: *"Are we making those practices as visible and salient as they deserve and need to be?"* This volume addresses these weighty questions in the hope that readers ask themselves and their stakeholders those same questions, and, if the answer is that more personal sustainability action is required (Kanashiro, Rands, & Starik, 2020; Parodi & Tamm, 2019), then, let's all get to it!

WHY WE INITIATED THIS EXPLORATION

Several decades ago, as a graduate student at a prominent U.S. university, one of the co-editors of this volume, was listening to his Energy Policy master's course instructor in class discussing a number of academic and professional energy events that the instructor either had attended or soon would be attending. This distinguished, well-respected scholar, who had written several energy policy books, listed perhaps a half-dozen or more such events, some of which included him as a featured speaker on various aspects of energy sustainability, and all of which were strewn widely around the world, requiring dozens of airline flights, tens of thousands of miles of air travel, and many tons of CO_2 emissions, the main greenhouse gas responsible for our global climate crises. The student thought to himself and would later discuss with a fellow student in the class whether or not their energy-conscious professor should be, as one of them described it, "flying all around the world, trying to save it," and whether we too, if we had the opportunity, would do the same. Several hundred flights and multiple decades later, that student, now a faculty member, and co-editor of this volume, is still asking himself that same vexing question, as well as many others, related to sustainability consistency. Cumulatively, he has flown "around the world" several dozen times, either for business or for pleasure, which he has finally, after several decades, stopped (except for emergencies). Some of us may have answered the above sustainability question in the negative, choosing to fly only when absolutely necessary. Apparently, and thankfully, many academics and others are also asking themselves similar consistency questions in the present day (Ciers, Mandic, Toth, & Veld, 2019). Others have justified their flights (and other problematic unsustainable behaviors) in one manner or another and/or have purchased carbon offsets. Still others have likely decided to either stop or keep asking that and related questions and may never answer them. Given that the average Chicago-to-Frankfurt roundtrip flight (before electric or biofuel technologies become widespread) generates 2.3 tons of CO_2 (myclimate, 2020) or more than one-tenth the average American's or one-fifth the average European's annual carbon footprint, that is a significant outsized carbon impact for a single flight. Of course, in addition to airline flights, this co-editor has also himself practiced and witnessed in some of his own and his former sustainability professors' and current sustainability colleagues' behaviors that might charitably be described as "questionable" from a sustainability perspective. Such behaviors have included owning and operating one or more very large vehicles, including sport utility vehicles, recreational vehicles, and motorized boats, and living in one or more very large homes significant distances away from our or their respective universities or other work locations. On the other hand, we, they, and others, may have lived in solar (active or passive) or otherwise energy-efficient homes, driven small electric automobiles or used alternative forms of transportation, and/or have assisted in setting up and running local food banks or other community-oriented activities, among many other laudable (and sustainability-oriented) actions.

The flying question was one seed among many others that relate to the same general question about becoming and being a sustainability academic: To what extent and how should faculty members, especially those who verbally promote sustainability values in their teaching and research, put those values into practice and communicate with their stakeholders about both those values and those practices? As mentioned above, none of us is always consistent in "practicing what we preach," whether on sustainability or any other topic. However, the normative nature of sustainability and its implicit question, *"How should we live, both environmentally and socio-economically, for the benefit of current and future generations and the rest of the planet?"* seems to have embedded the need for sustainability faculty to be active generally, if not consistently, on several levels of sustainability (Starik & Kanashiro, 2020).

So, let's stipulate that none of us is perfect, whether we are teachers, researchers, practitioners, students, or any of the dozens of other roles we play in society throughout our lives. We, the co-editors, were not seeking perfection in this volume's chapters (we like the saying, "Progress not perfection"), but we did want to illuminate faculty personal sustainability practices and advance accountability regarding those practices. Most of us would likely agree that, if at all possible, teachers should generally practice what they preach (Bergart & Simon, 2004), but since many of us are aspirational, we should also expect that we will probably not practice everything we preach all the time. However, shouldn't that fact be admitted and each of us be open and accountable for the gap between what we teach, research, or otherwise espouse, on the one hand, and what we really practice, on the other (Ruff, 2019)? This consistency between faculty teachings and faculty practice is sometimes termed congruency (Swennen, Lunenberg, & Korthagen, 2008), and a number of suggestions have been forwarded to advance teacher congruency, with some of those described in this chapter and book. For example, teachers who actively engage with their students and who believe that learning can be generated from both past successes and past failures can share both of these sets of experiences with their students (Mihai, 2014), so that students not only learn these sustainability lessons, but are more motivated to put them substantively into practice in their own lives and careers.

SUSTAINABILITY CHALLENGES AND OPPORTUNITIES IN OUR TIME

Questioning one's own consistency is one way to explore the sustainability preaching–practicing dichotomy, but another is to pay attention to and possibly emulate others who have apparently resolved, at least in part, that dilemma successfully over time. In one co-editor's case, numerous and sometimes well-known authors, activists, and teachers helped influence and inform this sustainability consistency exercise, including (in alphabetical order by last name): T.C. Boyle, Lester Brown, Robert Bullard, Archie B. Carroll, Cesar Chavez, Arthur C. Clarke, Barry Commoner, Walter H. Corson, Jared Diamond, Paul Ehrlich, John Kenneth Galbraith, Bill Gates, Mahatma Ghandi, Jane Goodall, Al Gore, Asterios G. Kefalas, John Kerry, Ken Keyes, Jr., Martin Luther King, Naomi Klein, Winona LaDuke, Frances Moore Lappe, Annie Leonard, Amory Lovins, Wangari Maathai, Donella H. Meadows, Bill McKibben, Chico Mendez, George Monbiot, Barrie Pittock, Robert Redford, Kim Stanley Robinson, Carl Sagan, Paul Shrivastava, David Suzuki, Jessica Yinka Thomas, Studs Terkel, Neil de Grasse Tyson, Kurt Vonnegut, and Ken Saro Wiwa, among many others. Additionally, three visual but distant examples decades apart were especially both salient and serendipitous for this co-editor's involvement in this project. The first was his accidental stumbling onto a YouTube video featuring Dr. Peter Kalmus, and subsequently purchasing and viewing the video entitled *Being the Change: A New Kind of Climate Documentary* (Grandelis & Davis, 2018), which tells the story of a N.A.S.A. atmospheric scientist (Dr. Kalmus) who not only addressed the climate crisis at work but also reorganized his and his family's personal lives to take whatever climate and sustainability actions they could at that level, including using biodiesel made from recycled restaurant cooking oil in his aging family car, beekeeping, gardening, bicycling, meditating, deciding with his wife to limit their family size, and actively participating in the Citizens' Climate Lobby and his local community climate projects. The second set of actors who appeared to be walking their sustainability talk (and then some) were Los Angeles (Studio City) neighbors Ed Begley, Jr. (whom the co-editor met at a Washington, D.C. sustainability conference) and Dr. Bill Nye (yes, "the Science Guy"), each of whom had their own sustainability-related television shows in the U.S. a few years ago and whose modest abodes were located in the same L.A. neighborhood. On several episodes of Begley's show called "Living with Ed," they engaged in a friendly competition over which one of their homes was more sustainable. While who won the competition was not important, what was important and impressive was their mutually amiable sustainability contest and their skill, imagination, and determination in installing the materials that were highly energy efficient, water efficient, and non-toxic, at reasonable cost in time and money (Schwartz, 2008). Yet a third positive inspiration in consistent sustainability for this co-editor is Dr. Doug McKenzie-Mohr, who developed the concept of Community-Based Social Marketing, which argues for the use of social psychology strategies, such as prompts, commitments, pilot testing, community orientations, and evaluation, to foster—that is, disseminate—sustainable behaviors related to agriculture, energy, transportation, waste, and water. In his work, he uses his own personal, household, and community examples, such as whether to compost in the Canadian winter and which of several transportation modes he chooses to get to work, to illustrate this sustainability diffusion, and does so through books (McKenzie-Mohr, 2011), conference speaking, online and in-person workshops (attended by this co-editor), and a website (cbsm.com) with hundreds of examples of his approach. Kalmus, Begley, Jr., Nye, and McKenzie-Mohr, each a "public educator" in his own way, were inspiringly consistent in their personal and professional sustainability behaviors and provided much-needed multi-decade boosts to this co-editor's efforts to follow, as much as possible, their inspiring "walking the talk" real-world examples.

Continuing on the subject of sustainability challenges and opportunities, one can imagine that a collection of chapters on the topic of faculty personal sustainability would likely include a wide diversity of subtopics and themes, and, at least on this point, the volume you are reading will likely not disappoint. Some of that diversity was by design, as the co-editors ensured that the call for proposal submissions included multiple possible subjects related to both environmental and socio-economic sustainability and to the several academic functions of research, teaching, and service, as well as the rest of our lives. We also wanted to ensure that faculty and other scholars and practitioners associated with any of a number of divisions and interest groups within several academic associations representing numerous world geographic regions, especially the Academy of Management's Organizations and the Natural Environment (O.N.E.) and Social Issues in Management (S.I.M.) divisions, knew their submissions would be welcomed for consideration. The topic diversity of the submissions we received, though, might have been either enhanced or hindered by the fact that, throughout most of the year 2020 and the initial weeks of 2021, several major global developments were occurring as these chapters were being curated, written, reviewed, rewritten, copyedited, structured, and compiled. The global COVID-19 pandemic was devastating numerous human populations around the planet, disrupting an incalculable number of lives and forcing numerous changes in how members of our species, including our contributors and their stakeholders, interacted with one another, whether in our homes, schools, communities, or workplaces, at social gatherings, and, of course, online. In addition to affecting our social discourse, the pandemic has also caused major economic disruptions, as widespread lockdowns and business closures meant significant adjustments for people in societies worldwide, worsened by the chaotic or non-existent distribution of vaccines. And, the world's richest but most unequal economy, the U.S., experienced both a major recession and a significantly contentious set of national and state elections, as well as a major upwelling of both political unrest and conflict and social justice-related events and responses, all of which had significant both national and international implications. Through all of this chaos and tumult, our contributors persevered and delivered an array of interviews and essays that offer a wide range of perspectives, experiences, concepts, and practices that we think readers will appreciate, consider, and employ in their own faculty personal sustainability plans and actions, at home, at work, at play, and on the way. In organizing these submissions, we identified several broad themes and have grouped them accordingly.

HOW THIS BOOK IS ORGANIZED

In our call for proposal submissions, we asked our faculty colleagues to consider their personal sustainability practices related to the 17 United Nations Sustainable Development Goals (SDGs), which set tangible social, environmental, and economic goals that, according to many world experts, need to

Table 1.1   *United Nations Sustainable Development Goals (UN SDGs)*

Goal no.	Goal description
1	End poverty in all its forms everywhere
2	End hunger, achieve food security and improved nutrition and promote sustainable agriculture
3	Ensure healthy lives and promote well-being for all at all ages
4	Ensure inclusive and equitable quality education and promote lifelong learning opportunities for all
5	Achieve gender equality and empower all women and girls
6	Ensure availability and sustainable management of water and sanitation for all
7	Ensure access to affordable, reliable, sustainable, and modern energy for all
8	Promote sustained, inclusive, and sustainable economic growth, full and productive employment and decent work for all
9	Build resilient infrastructure, promote inclusive and sustainable industrialization, and foster innovation
10	Reduce inequality within and among countries
11	Make cities and human settlements inclusive, safe, resilient, and sustainable
12	Ensure sustainable consumption and production patterns
13	Take urgent action to combat climate change and its impacts
14	Conserve and sustainably use the oceans, seas, and marine resources for sustainable development
15	Protect, restore, and promote sustainable use of terrestrial ecosystems, sustainably manage forests, combat desertification, and halt and reverse land degradation and halt biodiversity loss
16	Promote peaceful and inclusive societies for sustainable development, provide access to justice for all, and build effective, accountable, and inclusive institutions at all levels
17	Strengthen the means of implementation and revitalize the global partnership for sustainable development

Source: U.N. Department of Economic and Social Affairs, 2020.

be met by the year 2030 in the areas of climate action, poverty, hunger, infrastructure, health, and other pressing global issues (see Table 1.1). The United Nations calls on individuals, organizations, and governments to act together, globally and locally, to address these serious concerns. The urgency for each of us to act is even more necessary as the COVID-19 pandemic impacts, usually negatively, all of the 17 SDGs (UN/DESA, 2020).

This volume comprises essays and interviews by 36 authors representing 25 institutions, seven countries (United States, United Kingdom, Austria, Germany, France, Australia, and Romania), and at least six disciplines (Management, Economics, Finance, Physics, Biology, and Fine Arts). Each manuscript was evaluated following a double-blind peer observation and review process and revised at least twice. The result is a collection of diverse perspectives and inspiring insights on the authors' personal sustainability practices. We applaud the authors' accomplishments and continuous efforts; we are humbled by their trust in us and in this project, as the authors agreed to open up and share their personal sustainability plans, successes, failures, and frustrations. Finally, since the United Nations Sustainable Development Goals serve as one of the overarching themes for this book, we matched how each of the authors' practices may contribute to advancing one or more of the U.N. SDGs (see Table 1.2).

The book is divided into four parts. In Part I: Sustainability Practices in Action, the authors describe and reflect on how systems thinking contributes to sustainability actions for a more sustainable world. Robert Sroufe summarizes his personal sustainability activities, especially those related to buildings, using the ABCD (Awareness, Baseline Understanding, Creative Solutions, and Down to Action) planning process and offers insights on how his students and other stakeholders can scale their respective efforts more quickly and systematically than otherwise. Bernadette Roche argues that faculty need to adopt a teacher–scholar–practitioner action-oriented identity and aspire to have broader influence in multiple systems. Jimmy Y. Jia and Rick Dickinson introduce a methodology that highlights how an action in one subsystem impacts and ripples through adjacent subsystems. Kevin D. Carlson and John H. Grant develop a model tied to the UN SDGs to guide business college (and their students') transition towards sustainable development actions. Finally, Billy Friebele offers a practical application on how he merges sustainability practices with his artistic creations. We are grateful to Billy for allowing us to feature a photo of his sculpture, entitled "Disposable Empires", as the cover of our book.

In Part II: Internal/External Integration (Values to Action), the authors discuss opportunities and challenges as they attempt to leverage and connect their own personal sustainability values, visions, and beliefs to their sustainability practices, with the goal of transforming their respective households, organizations, communities, classrooms, and institutions. Shelley F. Mitchell interviews Paul Shrivastava to understand how personal sustainability values and actions inform his various leadership roles in academia, including his current position as Chief Sustainability Officer at Pennsylvania State University and his leadership in the Academy of Management as the lead co-founder of its Organizations and the Natural Environment (O.N.E.) division. Amy K. Townsend reflects on the need for both inner (values) and outer (action) dimensions of multiple personal sustainability practices over decades, both within and outside of sustainability-related organizations. Ralph Meima shares his professional and personal sustainability journey, while questioning whether current managerial teaching and tools can adequately address sustainability crises. Bruce Paton relates his experience with public service and challenges his students (and our readers) to "think like a city" when developing

Table 1.2   *Authors' contributions to the United Nations Sustainable Development Goals*

Author/s	Summary	UN SDGs
Part I: Sustainability Practices in Action		
Sroufe	Application of ABCD (Awareness, Baseline, Create a Vision, and Down to Action) planning methodology to scale efforts and more quickly implement sustainable solutions.	7, 9, 11, 12, 13
Roche	Reflection on why faculty should adopt a teacher–scholar–practitioner identity and aspire to have broader influence including impacts on public and cultural spheres.	1–17, in general
Jia & Dickinson	Presentation of a methodology to evaluate how mitigating actions taken in one subsystem can impact and ripple through adjacent subsystems.	7, 12, 13
Carlson & Grant	Application of a proposed model framework ("spinning top") to describe business college transition and progress towards sustainability linked to achievement of the UN SDGs.	1–17, in general
Friebele	Merging sustainability practices with artistic creation.	9, 11, 12, 13, 14, 15, 17
Part II: Internal and External Integration (Values to Action)		
Mitchell	Interview with Paul Shrivastava and his journey building institutional capacity for sustainability.	1–17, in general
Townsend	Reflection on the balance between doing sustainability (outer world) and being sustainable (inner world).	7, 12, 13, 15, 16
Meima	Evaluation of the limits of sustainability-oriented education and research in meeting climate change goals and proposal of practical ideas for managers to effectively address sustainability challenges.	7, 8, 13
Paton	Discussion on how to leverage personal civic activism and public service to explain key challenges and changes in achieving sustainability.	1, 7, 13, 17
Stone	Reflection on teaching sustainability values and how to model personal sustainability behaviors to students.	3, 7, 8, 12, 13
Miller	Reflection on how building a sustainable cabin for personal use provided teaching insights on the tradeoffs between beauty, interconnectedness, and utility provided by nature.	6, 7, 9, 11, 12, 15
Part III: Curriculum Development in Sustainability Education		
Edwards & Stubbs	Creating space for Earth System talk by walking a radically reflexive path in sustainability education curricula.	1, 2, 5, 6, 7, 12, 13, 16

Author/s	Summary	UN SDGs
Heuer	Discussion of reasons why teaching and research in business ethics require risk taking and innovative theories and strategies.	7, 12, 13, 14, 16
Cocke, Gentsch, Hefley, & Reichert	Development of a new (undergraduate) course with a focus on the UN SDGs.	1–17, in general
Nelson & Ionescu	Discussion of a course development with a focus on changes and adaptation in response to the climate crisis.	12, 13
Part IV: Faculty Personal Sustainability as Social Movement		
Kanashiro	Interview with Jessica Thomas and her leadership founding a network of over 2,000 scholars engaged in leveraging the power of business for good.	8, 10, 11, 12, 17
Obara, Klangboonkrong, Chapman, & Frank	Reflection on the role of educators as key drivers in changing paradigms in the pursuit of a more sustainable future.	1–17, in general
Delmestri, Etchanchu, Bothello, Habersang, Huerter O, & Schüßler	Launching of a new social movement "Organization Scientists for Future" (OS4Future) comprised of management scholars who seek to act on the climate emergency.	11, 12, 13
Venkatesan	Discussion of how sustainability has shaped the author's role in teaching economics and civic activism, including the founding of a non-profit organization focused on sustainability education and grassroot activism.	6, 12, 13, 14

sustainability solutions. Thomas E. Stone suggests that valuing one's time is the most important action related to sustainability that we can instill in our students and describes how he models his personal choices (including not owning a smartphone) to describe the concept and importance of time in the classroom. Van V. Miller walks us through the building process of his backwoods cabin to explain how his sustainability endeavors are anchored in three main concepts: beauty, interconnectedness, and utility, contrasting conservation and preservation.

Part III: Curriculum Development in Sustainability Education provides reflections and examples on how authors developed new curricula around topics related to the UN SDGs. Melissa Edwards and Wendy Stubbs advance a radically reflexive approach calling for holistic and systemic views that challenge and question assumptions and values. Similarly, Mark Heuer challenges faculty members to step outside an institutionalized mindset that rewards incremental reactive advances and urges colleagues to be more proactive and take greater risks in business ethics research, teaching, and service. The following two sets of authors offer practical suggestions and examples of curriculum development. Gary Cocke, Joanna Gentsch, William E. Hefley, and

Carolyn Reichert describe a new course that covers the breadth and the inter-relationships among the UN SDGs and report that students felt energized and engaged to act on their learnings. Dave Nelson and George Ionescu explain how to integrate course design, processes, and outcomes to teach corporate social responsibility.

Finally, Part IV: Faculty Personal Sustainability as Social Movement reflects on how faculty can launch, join, and/or lead social movements to advance sustainability goals. Patricia Kanashiro, one of this volume's co-editors, interviews Jessica Thomas, a co-founder of B Academics, a network of faculty interested in understanding how businesses can be a force for good and how she encourages interactions between B Corps and her sustainable entrepreneurship students. Louise Obara, Te Klangboonkrong, Gary Chapman, and Regina Frank reflect on the role of educators leading the development of sustainability policies and practices. Giuseppe Delmestri, Helen Etchanchu, Joel Bothello, Stefanie Habersang, Gabriela Gutierrez Huerter O, and Elke Schüßler report on their initiative founding of Organization Scientists for Future (OS4Future), a worldwide movement of scholars who seek to act on the climate emergency. Finally, Madhavi Venkatesan describes her experience as an academic economist teaching sustainability and as a founder of a non-profit organization focused on sustainability education and grassroots activism.

OBSERVATIONS, AND LESSONS LEARNED

The co-editors believe our contributors have provided a significant impetus regarding why we sustainability faculty need to walk our talk and how we might consider doing so more substantively, frequently, and effectively. We were heartened and inspired by the fact that so many of our colleagues were engaged in such a wide variety of personal sustainability practices and were apparently very effective in their efforts. We have identified several future opportunities that were not fully developed in this volume, and which were likely not emphasized enough in our initial call for contributions. First, most of our authors and interviewees focused primarily on environmental sustainability, with a few notable exceptions, and either did not mention or did not focus on socio-economic sustainability, in particular UN SDGs #1 (no poverty), #2 (no hunger), #3 (good health and well-being), #4 (quality education), and #5 (gender equality). In addition to environmental sustainability, we encourage greater attention to socio-economic issues, such as immigration, racial and gender equity, and wealth and income gaps.

Second, several concepts that are often associated with sustainability did not seem to receive as much attention as we expected in our contributors' chapters, though there were several much-appreciated contributor exceptions. Those topics included personal health, spirituality, biodiversity, permaculture, political involvement, voluntarism, and numerous aspects of management (including leadership, planning, information and measurement systems, policy, circular economy, supply chain, and investment). The co-editors take responsibility for not emphasizing these important sustainability aspects sufficiently in the call, and we hope to signal to readers that much more could be explored on the overall theme of this volume than we were able to include within it.

Third, sustainability in developing countries present their own opportunities and challenges, especially focused on socio-economic problems (Jamali & Keram, 2016), which was admittedly not the focus of this volume. The reasons why developing country faculty personal sustainability practices likely diverge from those in developed countries vary from restricted access to resources (e.g., personal vehicles, larger homes, mega stores, natural resources) to limited access to opportunities, such as travel and free time. Nonetheless, full-time academics in developing countries often command at least middle-class incomes, allowing some to adopt middle-class and sustainable consumption patterns, regarding personal vehicles, homes, and home appliances and furnishings.

So, what did the co-editors learn during this project that we want to share with our readers? First, we learned that walking the sustainability talk is a popular topic among sustainability academics and that many of our colleagues are at least privately (and some publicly) asking how consistent they are in practicing what they are preaching about sustainability. Second, we were heartened to know that so many of us are actively engaged in a wide range of sustainability practices, from micro to macro levels of action and discourse (Starik & Kanashiro, 2020).

Third, we are aware that any of several barriers may exist regarding why some faculty are hesitant to share their personal sustainability values and actions with their stakeholders, including their students. These hindrances may include a significant sense of humility about sharing personal information of any kind (i.e., what is "personal" is "private" for some of our colleagues), or a feeling that faculty are not expert or confident enough to share their personal sustainability practices with others. These are understandable potential obstacles but also create "out-of-site, out-of-mind" problems, so faculty are advised to begin with a moderate approach to the topic, so that they share some of their personal sustainability perspectives with some of their stakeholders, to the extent that both faculty members and their stakeholders feel comfortable regarding the selected sustainability topics, amount, intensity, and frequency of that sharing.

Finally, we sensed that most of us have a long way to go to be as consistent as we think we should be in walking our sustainability talk and that many of us are interested in sharing our experiences on that topic, all with the hope that, in the words of the new U.S. President Joe Biden, "we'll lead, not merely by the example of our power, but by the power of our example" (Biden, 2021).

NOTE

1.   We were deeply inspired by the following words of noted Brazilian teacher and philosopher, Paulo Freire: "I cannot be a teacher without exposing who I am. Without revealing, either reluctantly, or with simplicity, the way I relate to the world, how I think politically. I cannot escape being evaluated by the students, and the way they evaluate me is of significance for my modus operandi as a teacher. As a consequence, one of my major preoccupations is the approximation between what I say and what I do, between what I seem to be and what I am actually becoming" (Freire, 1996, p. 87).

REFERENCES

Bergart, A.M. & Simon, S.R. (2004). Practicing what we preach: Creating groups for ourselves. *Social Work with Groups*, 27(4): 17–30.

Biden, J. (2021). *Inaugural Address by President Joseph R. Biden, Jr.* [transcript]. The White House, January 21, at https://www.whitehouse.gov/briefing-room/speeches -remarks/2021/01/20/inaugural-address-by-president-joseph-r-biden-jr/ (accessed May 22, 2021).

Ciers, J., Mandic, A., Toth, L.D., & Veld, G.O. (2019). Carbon footprint of academic air travel: A case study in Switzerland. *Sustainability*, 11(1): 80, https://doi.org/10 .3390/su11010080.

Freire, P. (1996). *Pedagogy of freedom: Ethics, democracy, and civic courage.* Lanham, M.D.: Rowman & Littlefield.

Grandelis, M. & Davis, D. (2018) *Being the change: A new kind of climate documentary.* Gabriola, B.C.: New Society Publishers.

Haraway, D. (2016). *Staying with the trouble: Making kin in the Chthulucene.* Durham, N.C.: Duke University Press.

Jamali, D. & Karam, C. (2016). Corporate social responsibility in developing countries as an emerging field of study. *International Journal of Management Review*, 20(1): 32–61, https://doi.org/10.1111/ijmr.12112.

Kanashiro, P., Rands, G., & Starik, M. (2020). Walking the sustainability talk: If not us, who? If not now, when? *Journal of Management Education*, https://doi.org/10 .1177/1052562920937423.

McKenzie-Mohr, D. (2011). *Fostering sustainable behavior: An introduction to community-based social marketing.* Gabriola, B.C.: New Society Publishers.

Mihai, A. (2014). Seven things teachers preach but fail to practise. *The Educationalist*, November 11, at https://educationalist.eu/seven-things-teachers-preach-but-fail-to -practise-9cb71f0d56c8 2) (accessed May 22, 2021).

myclimate (2020). At http://www.myclimate.org (accessed September 1, 2020).

Parodi, O., & Tamm, K. (Eds.) (2019). *Personal sustainability: Exploring the far side of sustainable development.* Abingdon: Routledge.

Ruff, T. (2019). When teachers don't practice what they preach. *Lion's Roar*, at https:// www.lionsroar.com/when-teachers-dont-practice-what-they-preach/ (accessed August 2, 2020).

Schwartz, N. (2008). Ed Begley, Bill Nye wage friendly eco-war. ABC News, July 10, at https://abcnews.go.com/Technology/story?id=5353371&page=1 (accessed January 28, 2021).

Starik, M. & Kanashiro, P. (2020). Advancing a multi-level sustainability management theory. In Wasieleski, D.M. & Weber, J. (Eds.), *Sustainability (Business & Society 360, Vol. 4)*. Bingley: Emerald Publishing, 17–42, https://doi.org/10.1108/S2514-175920200000004003.

Swennen, A., Lunenberg, M., & Korthagen, F. (2008). Preach what you teach! Teacher educators and congruent teaching. *Teachers and Teaching: Theory and Practice*, 14(5–6): 531–542, https://doi.org/10.1080/13540600802571387.

UN/DESA United Nations Department of Economic and Social Affairs (2020). *UN/DESA Policy Brief #81: Impact of COVID-19 on SDG Progress: A statistical perspective*, August 27, at https://www.un.org/development/desa/dpad/publication/un-desa-policy-brief-81-impact-of-covid-19-on-sdg-progress-a-statistical-perspective/ (accessed January 27, 2021).

Wallace-Wells, D. (2020). *The uninhabitable Earth: Life after warming*. New York, N.Y.: Tim Duggan Books.

2. Why focus on faculty personal sustainability?

Mark Starik, Patricia Kanashiro, and Gordon Rands

In our collective experience, we have witnessed both ourselves and most of our colleagues who teach, research, and otherwise practice sustainability apparently doing much more talking and writing about sustainability than visibly practicing it. We say "apparently" and "visibly" because we and our colleagues have not generally made those sustainability actions (or their outcomes) obvious to one another, or, to our knowledge, to many of our and their stakeholders. We sustainability academics may be engaged in many different types of behavior to reduce our respective carbon footprints, for example, but how often have any of us discussed our actions related to energy efficiency, renewable energy, or more broadly, sustainable consumption, whether we are in the classroom, in a faculty meeting, at a conference, or at a ball game? The co-authors of this chapter have done so more frequently over time, but they have seldom heard their colleagues do so in public settings. Some of us might teach aspects of each of these topics and identify the benefits of practicing sustainability when we teach, yet how many of us mention our own personal sustainability actions, or, just as importantly, our lack of actions? Exceptions do exist, and the authors of chapters in this volume highlight some of these exceptions with the intent of encouraging readers of this volume to do so more frequently, substantively, and visibly.

Our hope is that we, the co-authors of this chapter, this volume's authors, and all of our readers (at least those who are or will become academics), both practice our sustainability talking and writing more substantively and that we communicate these efforts and their results more effectively, so that our collective positive impact helps spark a threshold advance in sustainability thinking and doing. Ideally, faculty may have the opportunity to develop a "sense of community" (Chavis & Pretty, 1999) of sharing sustainability practices throughout a course, a program, a department, a school, or a university (and its wider community).

REASONS WHY WE SUSTAINABILITY FACULTY NEED TO CONSIDER "WALKING OUR TALK"

Multiple reasons can be forwarded regarding why academics need to actively engage in sustainability and to communicate those actions (and their results) to faculty stakeholders (Kanashiro, Rands, & Starik, 2020). First, academics are privileged and presumed to have access to knowledge and time to reflect and share information, and, hopefully, sustainability faculty are more knowledgeable than the general public about the possibilities of, need for, and impact of personal sustainability actions. Faculty can draw on numerous and deep databases, provide examples and explanations, make important linkages, use specialized language, and develop expertise in either developing or utilizing course materials, such as textbooks, videos, online links, and slide presentations. Therefore, faculty, as "thought-and-action leaders," through their words and practices have the potential of living more sustainable lives, influencing many others, and engaging as many people as possible, which appears to be an ever-increasing imperative for the world's societies, in general (United Nations, 2019). Faculty, as teachers and researchers, play a vital role in our students' education, and younger or newer students may be especially open to suggestions from teachers and researchers. If the goal is to encourage those students and other stakeholders to adopt sustainable behaviors, academics may need to offer their own sustainability practices, experiences, and results as replicable models (Higgs & McMillan, 2010).

Second, faculty who have adopted various sustainability practices are obviously in a better position to inform their stakeholders about how those practices can be put in place most effectively and efficiently. Such "practicing what you preach" has the advantage of "knowing what you are talking about" (Seers, 2017). As one simple example, one of the co-authors enjoys cooking vegetarian organic chili in a system that uses reflected solar energy. The fact that he has done so numerous times, using either of two different systems, has allowed him both to demonstrate this technique to various stakeholders and to identify experts throughout his community network who give demonstrations of these and related solar cooking systems in his classes and to the general public (Solar Household Energy, 2019).

Third, faculty can serve as role models, primarily in the classroom (Rask & Bailey, 2010) but beyond the classroom as well, including throughout student careers (Erkut & Mokros, 1984). However, faculty promoting sustainability actions in their teaching may need to make that intention obvious by, among other ways, including those behaviors in student assignments and grades (Shepherd, 2008) and broader conversations. One of the co-authors knows several people who have either bought hybrid vehicles or rooftop solar panels due in part to his discussion of his experience owning these technologies, as well as many students who have expressed the desire to purchase such products after he has discussed his personal experiences in class.

Fourth, faculty are key stakeholders in universities, colleges, and other educational institutions. If these institutions and their societies are to advance toward the near universally stated goal of socio-economic and environmental sustainability, faculty themselves need to practice sustainability at least in their professional lives (Vargas, Mac-Lean, & Huge, 2019).

THE WHAT AND HOW OF COMMUNICATING SUSTAINABILITY ACADEMIC PRACTICES

Academics who study and teach sustainability topics, including a wide range of subjects that could also include policy, science, strategy, ethics, structure, systems, humanities, and many other organization-level topics, and who want to help foster sustainability behaviors among their stakeholders, including students, colleagues, and members of the general public likely have many such behaviors to share (The Minimalist Vegan, 2021) and we list a number of these possibilities later in this chapter. The maxim "start small—but start now" may be an appropriate decision rule in deciding which sustainability behaviors would best be adopted and shared. For instance, one of this chapter's co-authors started his personal sustainability communication with his students in one class by discussing with them his several years of mostly car-free experience, calculating and revealing the impacts of his mass transit trips versus his compact car rentals. That activity led to several student teams electing to develop class projects on the topic of sustainable transportation in different geographic areas.

Generally, faculty may want to start practicing sustainability more frequently and substantively by becoming aware of and examining their own everyday behaviors and noting the most obvious ways to promote socio-economic and environmental well-being. For faculty and other individuals in developed countries, two of the best ways to start to act more sustainably are to waste less and share more. Saving water, energy, and food from being wasted have multiple environmental benefits and can often be done at home, at work, at play, and on the way, and identifying and supporting worthy causes can have numerous socio-economic benefits (Charity Navigator, 2020; Green America, 2020). Given that faculty often assume leadership roles both within and outside the classroom, a prudent next step is advancing a personal sustainability plan that includes the saving and sharing practices mentioned above, thinking in terms of participating in a "circular economy" (Ellen MacArthur Foundation, n/d) and sharing one's sustainability plan with stakeholders as broadly and strategically as possible (Fischhoff, 2020; Tiuttu, 2019). For those faculty who are interested in applying sustainability in the form of the United Nations Sustainable Development Goals, B1G1 offers an online tool with suggestions (B1G1, 2020), the United Nations has published "The Lazy Person's Guide to Saving the World" with actions that individuals can take at home, at work, and in the community (United Nations, 2019) and we have elsewhere suggested ways to leverage behaviors related to the SDGs within courses (Kanashiro, Rands, & Starik, 2020).

POSSIBLE CONSIDERATIONS IN IMPLEMENTING FACULTY PERSONAL SUSTAINABILITY

Once again, the aim of this volume is not sustainability perfection at any level, even if such a concept could be defined. No matter how strongly any of us believes in sustainability as a desirable set of outcomes, achieving perfection is not the point. Rather, we are living in the "real world," which is chock-full of contradictions, uncertainties, errors, circumstances, conflicts, and limits. We are aware that some recent sustainability research has termed these kinds of human decision and action imperfections as "paradox," which we acknowledge as real-world flaws and legitimate concerns (Hahn et al., 2018). Applied to faculty personal sustainability situations, the locations of our and our partner's workplaces, the availability of sustainable housing and transportation, the existence of healthy food markets, the location of our children's schools and other activities, the geographic locations of other family members and many other variables can place constraints on our personal and household decision-making, including our own sustainability decision-making. The co-authors of this chapter try to practice (and recommend to their colleagues) any of several suggested approaches to implement and promote sustainability practices in our imperfect world, including "satisficing," equifinality, and kaizen. In this context, satisficing is considering a range of possible sustainability actions and outcomes and selecting one or more of those that meet accepted criteria, while equifinality is choosing one or more sustainability actions among several that would achieve the same goal. Finally, kaizen, or never-ending improvement, related to sustainability, is identifying solutions through ongoing innovation, problem-solving, and persistence (Staton & Winokur, 2013).

SOME POSSIBLE CATEGORIES AND EXAMPLES OF FACULTY PERSONAL SUSTAINABILITY ACTIONS

Faculty who are interested in engaging in personal sustainability actions and communicating such practices to their respective stakeholders have a wide range of choices from which to select, and these might be classified into two groups of actions, depending on where those actions could be taken—either in the faculty member's home or community or in their workplace or social setting.

Home and Community Faculty Sustainability Practices

Personal sustainability actions in this category are those aimed at directly reducing a person's negative and increasing their positive sustainability impacts at or near their own residence or while using their own possessions, such as engaging in organic gardening, recycling and composting, conserving home electricity or fuel in their vehicle, including by using renewable energy technologies, an electric vehicle, ride-sharing, and/or mass transit. Other possible actions include buying fair trade and organic products, donating to a food bank, frequenting a farmer's market, using videoconferencing to supplement or reduce the need for personal visits, donating to local causes, and marching locally for social and/or climate justice. This set of practices could include calling, writing, or visiting organizations that provide their household or community with products or services to encourage those organizations to change their practices so that fewer negative and more positive social and environmental impacts would be experienced in the home and/or community settings. This category could also include faculty and community members informing one another and their respective stakeholders about actions that can increase household sustainability in the community—that is, making such actions salient rather than stealthy.

Workplace and Societal Faculty Sustainability Practices

Personal sustainability actions in this category are similar to those in the Home and Community category above, but they occur in the workplace (such as on a university campus or other work-related settings) or in public settings beyond one's own community, such as university suppliers' or customers' worksites and conference hotels. Actions such as recycling, choosing vegetarian or "fair trade" options in the cafeteria, bringing a water bottle to the worksite in order to avoid the use of plastic cups, using double-sided printing, and turning computers and peripherals off when not in use all could fall in this category. Behaviors in this segment involve taking action to try to encourage the organizations and individuals with which a faculty member interacts in carrying out job-related duties—primarily their employer but also suppliers, business customers, service providers, such as hotels, and trade and professional associations—to alter their practices, products, and services, so that the faculty member and other employees or members of the profession can act more sustainably while conducting business on these organizations' premises or using their products or services. Advocating for the creation of or participating in a green team or a campus sustainability committee, collecting data relating to one's employer's sustainability-related behaviors and impacts, or advocating that their professional organizations take steps to green their conferences could fit within this segment. Educating fellow faculty members, staff, students, and similar stakeholders about sustainability-related workplace behaviors, or about the negative impacts that they are experiencing, could also be classified into this category. This grouping could also include behaviors that involve taking action to try to encourage the organizations and individuals with which the faculty member interacts in public non-work-related settings to sustainably alter their practices, products, and services. Examples could include various means of giving feedback and advocating for change, such as leaving customer suggestions and complaints, writing to customer and public affairs departments, contacting government officials and regulatory agencies, writing letters to the editors of local newspapers and other media, investing in a socially responsible manner, and engaging in various kinds of sustainability-related demonstrations and political party, referenda, and candidate campaigns. It could also include informing others about these concerns and possible actions and attempting to encourage their involvement in such activism, as well.

In this chapter, we have suggested why faculty should engage in personal sustainability behavior, to whom we should communicate those behaviors, how we can try to implement such behaviors, where such behaviors can be practiced, and what they might include. We have not directly addressed the question of when we should engage in such behaviors, but we believe that the severity of the multiple sustainability crises we face make the answer to that question obvious: we must do so now!

In summary, we encourage faculty to continually assess and find ways to practice sustainability and to communicate our own sustainability practices to our students, colleagues, and other stakeholders. We leave you with a quote attributed to Dr. Jane Goodall in 2010 (EcoWatch, 2021) to encourage ourselves, our colleagues, and our readers to consistently walk our sustainability talk: "You cannot get through a single day without having an impact on the world around you. What you do makes a difference, and you have to decide what kind of difference you want to make."

REFERENCES

B1G1. (2020). Sustainable Development Goals – A guide for businesses. https://www.b1g1.com/businessforgood/sustainable-development-goals. Retrieved December 15, 2020.

Charity Navigator. (2020). https://www.charitynavigator.org. Retrieved December 12, 2020.

Chavis, D.M. & Pretty, G.M.H. (1999). Sense of community: Advances in measurement and application. *Journal of Community Psychology*, 27(6): 635–642. https://doi.org/10.1002/(SICI)1520-6629(199911)27:63.0.CO;2-F

EcoWatch. (2021). These Jane Goodall quotes will inspire you to save the world. https://www.ecowatch.com/jane-goodall-quotes-2361011467.html.   Retrieved January 30, 2021.

Ellen MacArthur Foundation. (n/d). Circular economy. https://www.ellenmacarthur foundation.org/. Retrieved January 26, 2021.

Erkut, S. & Mokros, J.R. (1984). Professors as models and mentors for college students. *American Educational Research Journal*, 21(2): 399–417. https://doi.org/10.2307/1162451

Fischhoff, M. (2020). How to motivate people toward sustainability, June 22. https://www.nbs.net/articles/how-to-motivate-people-toward-sustainability.   Retrieved December 20, 2020.

Green America. (2020). 10 habits of highly sustainable people. https://green.america.org/green-living/10-habits-highly-sustainable-people. Retrieved October 5, 2020.

Hahn, T., Figge, F., Pinske, J., & Preuss, L. (2018). A paradox perspective on corporate sustainability: Descriptive, instrumental, and normative aspects. *Journal of Business Ethics*, 148: 235–248. DOI: 10.1007/s10551-017-3587-2

Higgs, A.L. & McMillan, V.M. (2010). Teaching through modeling: Four schools' experiences in sustainability education. *The Journal of Environmental Education*, 38(1): 39–53. https://doi.org/10.3200/JOEE.38.1.39-53

Kanashiro, P., Rands, G.P., & Starik, M. (2020). Walking the sustainability talk: If not us, who? If not now, when? *Journal of Management Education*, 44(6): 822–851.

Rask, K.N. & Bailey, E.M. (2010). Are faculty role models? Evidence from major choice in an undergraduate institution. *The Journal of Economic Education*, 33(2): 99–124.

Seers, A. (2017). Management education in the emerging knowledge economy: Going beyond "Those who can do, Those who can't teach." *Academy of Management Learning and Education*, 6(4): 558–567. https://doi.org/10.5465/amle.2007.27694955

Shepherd, K. (2008). Higher education for sustainability: Seeking affective learning outcomes. *International Journal of Sustainability in Higher Education*, 9(1): 87–98. https://doi.org/10.1108/14676370810842201

Solar Household Energy. (n/d). 2019 Annual Report. http://www.she-inc.org/?page_id=407. Retrieved January 26, 2021.

Staton, G. & Winokur, K. (2013). 5 lessons from the Japanese "kaizen" approach to sustainability. Greenbiz. https://www.greenbiz.com/5-lessons-from-the-Japanese-"kaizen"-approach-to-sustainability. Retrieved December 21, 2020.

The Minimalist Vegan. (2021). 100+ tips to live a more sustainable lifestyle. https://theminimalistvegan.com/live-a-more-sustainable-lifestyle/. Retrieved January 27, 2021.

Tiuttu, T.A. (2019). Sustainable consumption and creating a personal sustainability action plan. 4Circularity, October 18. https://4circularity.com/sustainable-consumption-and-creating-a-personal-sustainability-action-plan/.   Retrieved December 10, 2020.

United Nations. (2019). *The lazy person's guide to saving the world*. https://www.un.org/sustainabledevelopment/takeaction/. Retrieved January 20, 2020.

United Nations Department of Economic and Social Affairs. (2019). *The future is now: Science for achieving sustainable development. Global Sustainable Development*

Report. https://sustainabledevelopment.un.org/content/24797DSDR_report-2019. Retrieved November 3, 2020.

Vargas, L., Mac-Lean, C., & Huge, J. (2019). The maturation process of incorporating sustainability in universities. *International Journal of Sustainability in Higher Education*, 20(3): 441–451. https://doi.org/10.1108/IJSHE-01-2019-0043

PART I

Sustainability practices in action

3. Design for the experience: a more sustainable future

Robert Sroufe

INTRODUCTION

We can all start with where we live and how we get around to walk the sustainability talk (Kanashiro, Rands, and Starik, 2020). How many of us have told our students or colleagues what we have done to advance sustainability in our professional or personal lives or calculated our ecological footprint (Starik, 2015)? Calculating an ecological footprint is a great place to start a shared understanding of opportunities (Edstrand, 2015). Measuring an ecological footprint is something we do with every MBA cohort of students. It creates awareness of the impacts we have on Earth and provides a baseline understanding of resource consumption and systems' carrying capacity. Measuring an ecological footprint is a tragic place to stop if we do not leverage this understanding with action and more sustainable business practices. It is also a way to get students and faculty to think about the food they eat, the homes they will live in someday, and transportation systems that will take them to and from work.

Within this chapter, I am excited to capture some of the main aspects of what I do regarding my sustainability efforts and how I explicitly communicate my learning in my teaching. To this end, I have found success in leveraging the investments made in my home and transportation to bring my learning into my business school as a learning lab and even to set goals of getting personal sustainability efforts to a net-zero energy footprint. My interests have progressed to trying to answer the question, "How can I design and deliver unique learning experiences students will remember years after the course or degree program?" This is one of the ways I'm contributing to sustainability efforts locally and globally. My published work in this area has spanned MBA program development, experiential courses, design competitions, and live project courses (Sroufe and Ramos, 2015). Balancing my scholarly work involves research interests in sustainable supply chains, environmental management systems, the use of CO_2 in decision-making, industry 4.0, and high-performance buildings.

The intersection of my desire to deliver meaningful course experiences based on my research interests has provided several fruitful outcomes as I helped deliver a new innovative MBA program. Over the last thirteen years, I have developed or helped develop nine new graduate courses (spanning strategic sustainability and models, applied sustainability for new projects, supply chain management, a sequence of live project consulting courses, and international study abroad) (Sroufe, Sivasubramaniam, Ramos, and Saiia, 2015). I have also developed an undergraduate cornerstone experiential course and an honors course on imagining a sustainable world and collaborated across colleges within my university to deliver this content.

When designing new courses and delivering learning modules for students, I get the most energy out of project-based courses with teams of students and project partners. I like working with physical products, projects, and having the ability to try to test things while also taking them into the classroom. In the following sections, I will describe this approach and use the planning process to outline the sections of my awareness, baseline understanding, creative solutions, and decisions leading to success at home and in the classroom. Based on this, I can review how my family and I have improved our home's energy efficiency and generate our own energy at home at a 40 percent cost of our utility company charges. We generate more energy than we need for the house, offloading the extra energy to an electric vehicle and driving that vehicle at a cost of 90 percent less than a combustion engine vehicle. We have purchased offsets from vendors such as Native Energy, Green Mountain Energy, Terrapass, the Carbon Fund, and WGL Energy for 100 percent of the other energy we consume and other modes of travel as we attempt to get to goals of "zero" and try to walk the talk of sustainability.

These efforts are meaningful as I teach within a specialized graduate program integrating sustainability across courses as an endowed chair. In this capacity, I'm able to design and develop new course offerings. I am grateful for the opportunity to be an academic. It has allowed me to test ideas, pursue interesting research, collaborate with thought leaders in the industry, and work with future business leaders to test ideas.

DESIGN FOR THE EXPERIENCE THAT ENERGIZES YOU

After years of applying it to consulting work within my classes, and while on sabbatical in Sweden, I further validated the use of an ABCD planning methodology. This methodology is a rigorous, evidence-based approach to defining what is and what is not sustainable. I was introduced to the ABCD planning process as part of The Natural Step (Robért et al., 2015; Broman and Robért, 2017), and have utilized it through its transition into the current framework for Strategic Sustainable Development (Missimer, 2015; Broman and Robért, 2017; Sroufe, 2018). It is a simplified approach, much like Deming's approach to quality management using Plan, Do, Check, Act. Regarding sustainability, it uses Awareness, Baseline, Create a Vision, and Down to Action to form the acronym ABCD, which describes the strategic planning process within the framework to demonstrate its simplicity and power in problem-solving.

Awareness

Early in my teaching of sustainability, I learned I did not want to market the problem but instead sell the opportunities, get people to experience something new, and challenge the status quo. To this end, I can talk about how food travels an average of 1,500 miles to our plates, and we throw 40 percent of it away, while over 13 million households are food insecure (United States Department of Agriculture, 2020). When I turn on the lights, I do not think about the energy systems behind the switch and how we waste 67 percent of all energy created in the US (Lawrence Livermore National Laboratory, 2020). These 150-year-old systems are opportunities for change and more effective investment. One can also bring up the fact that 81 percent of every gallon of gasoline poured into a combustion engine vehicle is wasted on heat and friction, with roughly 19 percent going toward the movement of the vehicle (Lovins, Lovins, and Hawkins, 2007). These are starting points for awareness, but I needed more to convince students that sustainability is a valid part of business opportunities. Further, sustainability does not physically involve hugging trees. It includes managing environmental impacts and making a good business case for a project or capital expenditure.

It seemed like my early years of creating sustainability courses were centered on going beyond frameworks and models to enable better decision-making to prove the validity of sustainability concepts, actions, and tools. The proof was often found in leveraging top-ranked companies and their actions. Corporate sustainability reporting helped immensely in these early days. I developed classes in which, every other week, we were out on site to visit a leading regional company engaged in sustainable business practices. This included green roofs, renewable energy, closed-loop systems, e-waste, water systems, manufacturing, and high-performance buildings. Student teams and I did live consulting courses as a core requirement of the curriculum to see what sustainability challenges were presenting themselves in regional and national companies. I was able to develop and deliver a core required course on Strategic Sustainability. The project courses and this course became anchors for student awareness of best sustainable business practices. I was also much more aware of my actions and wanted to tell a more engaging story about what we can all do to be more sustainable at work and at home. The goals were to reduce waste in our home, lower energy consumption, and reduce our ecological footprint, and I knew we had to demonstrate a return on these investments.

Given all that had been happening in the green building movement over the last twenty years, buildings, including my own home, provided a grounded (Strauss and Corbin, 1997) opportunity to reduce my family's ecological footprint. Our homes can also provide examples of the best practices discussed in the classroom. These practices can also be used in business schools and andragogy to ensure students become more aware of the cost savings and reduced ecological footprint they can have after graduation.

Buildings are essential for several reasons. They can be large consumers of energy, water, and holding places for waste streams. Now consider that people spend 90 percent of their time inside buildings (pre-pandemic), so if you are 50, that is 45 years of your life. These spaces provide ample opportunity to measure, manage, invest in, and learn from throughout our lives. We can have goals for these spaces and the transportation systems connected to them to have very low energy consumption. These spaces can even generate energy and become living buildings. High-performance buildings can and should be utilized for inspiration and awe in how efficient they can be while providing us with cleaner air outside and a healthier lifestyle.

I was at the United Nations in New York City for the release of the UN's Sustainable Development Goals (SDGs) in 2015. The timing of this was also right for coming back to Pittsburgh, sharing these, enabling student awareness of global goals. Awareness was a purposeful part of the design and delivery of classes centered on experiential learning with outputs aligned with the SDGs (United Nations, 2015). Goal #17 Partnerships was a natural fit for finding and developing the live project consulting classes and study abroad with field trips to businesses. In every class, consulting project, and a design competition for teams of MBAs to invest in green buildings options for our business school, the options for the renovation of the school, and even in-home renovations, students and I could find alignment with the UN SDGs as a way to see local actions as having global consequences.

My awareness of building best practices and our home's potential as a catalyst for change led to 50 percent reductions in energy consumption. Next, we installed our first photovoltaic system to generate a 25 percent portion of our remaining energy needs for personal sustainability efforts. The home was listed on a solar home tour for years. This helped to engage others about the cost-effective benefits of first reducing our energy consumption and then generating our energy, consuming that energy, and getting paid renewable energy credits for doing so. The system has a 12-year payback and 25-year life.

Baseline Understanding

Food, buildings, and transportation are systems we can access and experience. Donella Meadows, author of *Thinking in Systems* (2008), offers that "a system is an interconnected set of elements that is coherently organized in a way that achieves something." Buildings are complex, interconnected systems that produce a pattern of operational behavior over time. If we change the performance of the building envelope, it will impact the mechanical systems. If you address lighting in the building, your solution has a relationship between the windows and artificial lighting systems, which impacts the heating and cooling systems. Only addressing one element at a time can have significant unintended impacts on the building's function or purpose due to the complex interconnection of those elements. Within a systems context, we take a closer look at our own homes, university buildings, and offices.

When it comes to buildings, we should walk the talk within our homes and scale this learning in multiple ways outlined in this book and the other chapters. We can learn from the places where we live, work, and learn. During the COVID-19 pandemic, these typically disjointed places have become one, and healthy buildings are even more critical now (Allen and Macomber, 2020; Solberg and Akufo-Addo, 2020). The built environment can demonstrate sustainability best practices at home, work, and school. When thinking, "Why are buildings important?" we can model sustainability in our behaviors at home, provide learning about our business school within student projects, communicate results, show commitments to sustainability, share practices, and demonstrate impacts to multiple stakeholders.

The best approach to reducing energy consumption inside a building is by using passive house options and the Natural Order of Sustainability (Sroufe, Stevenson, and Eckenrode, 2019). It promotes a Passive first, Active second, and Renewable last strategy, which ensures that a building's systems are optimized for performance. Maximizing *Passive* strategies (i.e., insulation, envelope, air barriers, thermal bridges, shading, windows, and doors within your home or any building) first helps reduce loads for *Active* heating and cooling systems, thereby requiring smaller and more efficient active solutions for mechanical systems to ventilate, heat, and cool buildings. We were not able to do an entirely passive house retrofit on our own home. However, we did bring down our consumption by improving the insulation, envelope, windows, and doors and installing water-conserving fixtures and aerators, Energy Star equipment, and LED lighting. In simple terms, the building will be more comfortable and cheaper to heat and cool, and the air quality will be better due to passive improvements. As a final step, *Renewables* can be used to zero-out remaining energy consumption and carbon emissions. At this point, renewables are more affordable due to their decrease in size and lower first-costs.

At home

For our own home's baseline information, we used its square footage and utility bills to get annual consumption of electricity, natural gas, and water consumption. This baseline information allowed us to determine our energy use intensity (EUI) kBtu/sq. ft performance before making passive improvements. We called in a company who works with our utility company to do a no-cost blower door test, to pressurize the house to see how leaky it was. Due to these baseline metrics and passive efforts, we could take a 150 percent larger than average home in Pennsylvania and bring its energy consumption down to 50 percent of what an average home consumes while also decreasing our water consumption by 30 percent. We were able to do subsequent blower door tests of the house and see improvements in less air leaking. We could do all of this with a good return on investment that kept paying back each year as part of our lower utility bills. We also calculated our CO_2 emissions from our home. Furthermore we applied a social cost of carbon dioxide (SCC) (Nordhaus, 2017) to be able to show the 12-year financial payback on the photovoltaic system was feasible. That payback only gets better when we include the environmental impacts avoided and human health and productivity improvements in applying the SCC. These initial efforts were not enough to get us to a net-zero energy level, yet provided a proving ground for success, moving us toward a vision of being more sustainable, were flexible, and had an excellent ROI to build upon for future projects.

At the university

We started with baseline information for our business school's energy consumption, water consumption, and indoor air quality monitoring. This baseline information is a catalyst for a design for the experience (DftE) annual competition using the business school as a living lab for students and faculty (Sroufe, 2020). We were able to have companies come to our 169,000 square foot building and pressurize the entire building. The building is over 50 years old and represents over two million buildings in the US that are approximately the same size and age (US Energy Information Administration, 2020). To date, this is one of the largest, if not the most extensive, blower door tests in the US. What students learned in this building applies to millions of other buildings that will house their future offices throughout their careers.

As part of a design competition, students have an unlimited budget to invest in the building. They must apply a social cost of carbon dioxide (SCC) to GHG emission reductions and demonstrate the financial return, environmental impacts avoided, and social value created (Sroufe and Jernegan, 2020). Students can see how building investments save money and improve indoor air quality, health, and wellbeing (Allen and Macomber, 2020). The baseline understanding of a building's performance puts students on a path toward experiencing awe-inspiring opportunities, with ideas for how to get the building to net-zero energy and the insight that they can also do these same passive and active improvements in their own homes. Students get hands-on experience with indoor air quality, WELL buildings, and RESET air standards and can see the growing importance of high-performance buildings.

Creative Solutions: Can We Get to Big Goals?

Learning from the home, the annual building competition and environmental and social performance in returns on investment have spilled over into the live project consulting courses (Sroufe and Ramos, 2011). We use the same ABCD planning process in the live project engagements and challenge all teams to find a minimum of $100k in value for our project partners. In doing so, student teams map their recommendations into the UN SDGs and translate learning across courses and buildings. One example of alignment with the SDGs is from the World Green Building Council (WGBC, 2020) and how improving buildings aligns with the UN SDGs. Green buildings can apply to the following SDGs:

#3   good health and wellbeing
#7   affordable and clean energy
#8   decent work and economic growth
#9   industry innovation and infrastructure
#11 sustainable cities and communities
#12 responsible consumption and production
#13 climate action
#15 life on land
#17 partnerships for the goals.

Outcomes of the learning from our own homes and green buildings have led to new insights. Instead of traditional financial calculations, we can integrate environmental and social impacts and benefits using the social cost of carbon dioxide (SCC) to provide enhanced financial valuations in the form of Return on Integration (ROInt) and Integrated Rate of Return (IntRR) (Sroufe, 2018). Furthermore, instead of focusing exclusively on the NPV of a decision, we have proposed using Integrated Future Value (IntFV) as a more robust measure of the long-term, future implications of investments made today. This integrated bottom line (IBL) approach to decision-making provides a dynamic business performance approach (Sroufe, 2018).

An example of integrated bottom line payback is in a lighting retrofit to LED lights with a five-year payback (this depends on several variables glossed over here). When we include the environmental impacts avoided, it comes down to a three-and-a-half-year payback. Based on work with Ph.D. architecture students at Carnegie Mellon, when we include human health and productivity impacts, this social value can be ten times better than a single bottom line return, with a payback of around four months (Shrivastava, 2018).

Creative outcomes can and should include lower costs of owning and operating homes, vehicles, and business schools, with benefits beyond first-costs as we realize environmental impacts avoided and social value created. COVID-19 and HyFlex have made buildings and lower energy consumption and links to air quality even more salient (Perron and Gross, 2020).

DOWN TO ACTION – PERSONAL SUSTAINABILITY – TEACHING & RESEARCH

The DftE challenge is not to allow students to propose incremental changes or only cost efficiencies. Instead, we can challenge them to propose big, awe-inspiring solutions, i.e., zero waste, carbon neutrality, net-zero energy buildings, or living buildings (which produce more energy and water annually than they consume). Setting big goals for homes, buildings, and transportation also provides new opportunities for supporting the UN SDGs (United Nations, 2015). Some of the decisions and outcomes of this ABCD process include the following:

- Investment in a second solar energy system and purchase price agreement (because we wanted to test the option compared to owning another system) to get our home to net-zero energy, i.e., we can produce all the energy we need and target BIG goals of zero (Elkington, 2012). We can also go beyond producing what we need for the home. We added a second solar system, so it produces an extra two megawatts of energy a year. Why? So we can offload this extra energy to an electric vehicle. The second system's investment makes the home net positive. Our energy costs are 40 percent less than our utility company (a price of 8.7 cents per kWh with no taxes or transmission fees) and can provide energy to us using a local 100-year-old coal-based utility system. Our local generation plant is one of the top ten most polluting in the country, and we can now talk about not needing its energy and our ability to reduce emissions.
- We bought an electric vehicle leveraging federal tax credits, and local credits come in at just over $5k, resulting in a total cost of $30k. We renovated the house four years in advance of the car to bring down the energy needed using passive insulation options for the walls and flooring while also wiring the home to charge the car. Using self-generated energy reduces our costs to drive the car 10k miles a year for $175. Over average lifetimes of driving (i.e., 50 years), at $2k per year for gas and

maintenance, a consumer will spend ~$100k on mobility. Unfortunately, consumers overlook the waste equivalent to approximately $81k of that money over time (Lovins, Lovins, and Hawkins, 2007). Electric vehicles can cost 90 percent less to drive per mile, contribute less to outdoor air quality, and require less maintenance (Errity, 2020; Borlaug, Salisbury, Gerdes, and Muratori, 2020). These energy management improvements at home are energizing to talk about in class and relate to students' research and experiential learning opportunities.

- Other creative outcomes include integrating learning with andragogy and research. I have engaged students, alumni, professionals, university facilities personnel, and administrators with a bottom-up approach to influence change within campus buildings (Brinkhurst, Rose, Maurice, and Ackerman, 2011). The work on the home, live projects, and consulting have led to five books in the last six years on Sustainable Supply Chain Management, Integrated Management, and High-Performance Buildings.

- To date, we have completed close to 200 live consulting projects to facilitate deep learning (O'Brien and Sarkis, 2014). No two have been the same as we charge student teams to find creative IBL solutions for project partners and integrate this learning across the curriculum each semester. Deliverables to our project partners have ranged from $100k to upwards of $40M for one semester-long project.

- The Natural Order of Sustainability can be applied to homes, offices, university buildings, and student competitions to get to innovative solutions that reduce energy consumption and improve indoor air quality. We now have a dashboard in the business school so occupants can see real-time performance data on the building. At home, we have integrated our energy generation dashboard with HVAC (heating, ventilation, and air conditioning) controls (using a Google Nest thermostat). We have indoor air quality sensors. The first, a Foobot, measures: indoor particulate matter, VOCs (volatile organic compounds), CO_2, humidity, and temperature. We can then compare these data with local outdoor air quality monitors. We also have a separate AirThings Wave sensor in the basement recording VOCs, CO_2, humidity, temperature, barometric pressure, and radon levels. These systems and sensors can be managed from our smartphones and used in class to demonstrate personal sustainability practices.

- Listing the home on a local solar home tour enabled it to get listed on a national net-zero energy buildings registry by the energy and environmental building alliance (EEBA). Our second system's installation led to four other homeowners in our neighborhood asking us about the process. Subsequently, these neighbors installed rooftop solar systems within the next two months.

- We also consider our natural gas consumption and air travel on an annual basis to purchase carbon offsets. When available, we select Pennsylvania wind options to promote local energy generation options competing with traditional fossil fuels and fracking in the region. We use these offsets to reduce our energy consumption with net-zero energy goals for the home and most of our travel.
- Recognition of these efforts from external organizations has helped to validate my teaching. To this end, I have received a Page Prize, as well as Aspen Institute, Decision Sciences Innovative Teaching, Production and Operations Management Society Innovative Teaching, and university teaching awards for the integration of sustainability into courses and curriculum.

LESSONS AND CONTINUED LEARNING

This chapter has blended self-reflection on personal sustainability and how this translates into designing experiential learning opportunities for myself, faculty, and students. Writing it has allowed me to describe the DftE mindset when developing experiences for both myself and others to learn about sustainable business practices. DftE and ABCD planning have helped me understand, realize, and communicate personal sustainability practices. What I hope readers can take away from this chapter is optimism. The spaces in which we live and work and the vehicles on which we rely too much for mobility in the US provide dynamic sustainability opportunities aligned with the United Nations' SDGs. By taking a deeper dive into understanding buildings such as our homes, offices, and businesses, along with how we get to and from those spaces, we can personally walk the talk of sustainability (Kanashiro, Rands, and Starik, 2020).

We should also challenge our colleagues and other stakeholders to be more sustainable (Arevalo and Mitchell, 2017). There is not enough space in this chapter to take on all the limitations or the entrenched business-as-usual attitude resisting the sustainability paradigm. To my accounting, economics, and finance colleagues: what are you waiting for? We need your help in measuring, valuing, and managing emerging business practices to make the intangible tangible. When we work together, we lay an experience-based path to a more sustainable world. This path includes collaboration and action-oriented research regarding how we can scale efforts to more quickly be the kind of change we want to see in the world (Pathak, 2018).

REFERENCES

Allen, J., and Macomber, J. (2020). *Healthy buildings: How indoor spaces drive performance and productivity*. Cambridge, MA: Harvard University Press.

Arevalo, J., and Mitchell, S. (Eds.). (2017). *Handbook of sustainability in management education: In search of a multidisciplinary, innovative, and integrated approach*. Cheltenham, UK and Northampton, MA, USA: Edward Elgar Publishing.

Borlaug, B., Salisbury, S., Gerdes, M., and Muratori, M. (2020). Levelized cost of charging electric vehicles in the United States. *Joule*, 4(17), 1470–1485.

Brinkhurst, M., Rose, P., Maurice, G., and Ackerman, J. D. (2011). Achieving campus sustainability: Top-down, bottom-up, or neither? *International Journal of Sustainability in Higher Education*, 12(4), 338–354. doi: https://doi.org/10.1108/14676371111168269

Broman, G. I., and Robért, K. H. (2017). A framework for strategic sustainable development. *Journal of Cleaner Production*, 140, 17–31.

Edstrand, E. (2015). Making the invisible visible: How students make use of carbon footprint calculator in environmental education. *Learning, Media and Technology*, 41(2), 416–436. doi: https://doi.org/10.1080/17439884.2015.1032976

Elkington, J. (2012). *The Zeronauts: Breaking the sustainability barrier*. Abingdon: Routledge.

Errity, S. (2020). What is the cost of running an electric car? Driving Electric, accessed January 6, 2020 at https://www.drivingelectric.com/your-questions-answered/466/what-cost-running-electric-car

Kanashiro, P., Rands, G., and Starik, M. (2020). Walking the sustainability talk: If not us, who? If not now, when? *Journal of Management Education*, 44(6), 683–698.

Lawrence Livermore National Laboratory. (2020). Energy flow charts: Charting the complex relationships among energy, water, and carbon. *The 2020 Flow Chart*, accessed August 12, 2021 at https://flowcharts.llnl.gov/

Lovins, A., Lovins, H., and Hawkins, P. (2007). A roadmap for natural capitalism. *Harvard Business Review*, July–August; ISSN 0017-8012.

Meadows, D. (2008). *Thinking in systems: A primer*. White River Junction, VT: Greenleaf Publishing.

Missimer, M. (2015). Social sustainability within the framework for strategic sustainable development. Doctoral dissertation series No. 2015:09, Blekinge Institute of Technology, Karlskrona, Sweden.

Nordhaus, W. D. (2017). Revisiting the social cost of carbon. *Proceedings of the National Academy of Sciences*, 114(7), 1518–1523.

O'Brien, W., and Sarkis, J. (2014). The potential of community-based sustainability projects for deep learning initiatives. *Journal of Cleaner Production*, 62, 48–61.

Pathak, A. (2018). The myth of value-free teaching. *The Wire*, accessed June 7, 2021 at https://thewire.in/education/the-myth-of-value-neutral-teaching

Perron, J., and Gross, S. (2020). Amid COVID-19, don't ignore the links between poor air quality and public health. *PlanetPolicy*, Brookings, accessed June 7, 2021 at https://www.brookings.edu/blog/planetpolicy/2020/08/19/amid-covid-19-dont-ignore-the-links-between-poor-air-quality-and-public-health/

Robért, K.-H. et al. (2015). *Strategic leadership towards sustainability*. Blekinge Institute of Technology, Department of Strategic Sustainable Development, Karlskrona, Sweden.

Shrivastava, R. (2018). Integrating financial, environmental, and human capital – the triple bottom line – for high performance investments in the built environment. PhD Thesis, Carnegie Mellon University, Pittsburgh, PA.

Solberg, E., and Akufo-Addo, N. A. D. (2020). Why we cannot lose sight of the Sustainable Development Goals during coronavirus. *World Economic Forum*, accessed January 7, 2021 at https://www.weforum.org/agenda/2020/04/coronavirus -pandemic-effect-sdg-unprogress/

Sroufe, R. (2018). *Integrated management: How sustainability creates value for any business*. Bingley: Emerald Publishing.

Sroufe, R. (2020). Business schools as living labs: Advancing sustainability in management education. *Journal of Management Education*, 44(3), 726–765.

Sroufe, R., and Jernegan, L. (2020). Making the intangible tangible: Integrated management and the social cost of carbon. *Sustainability*, 4, 163–183.

Sroufe, R. and Ramos, D. (2011). MBA program trends and best practices in teaching sustainability: Live project courses. *Decision Sciences Journal of Innovative Education*, 9(3), 349–369.

Sroufe, R., and Ramos, D. P. (2015). Leveraging collaborative, thematic problem-based learning to integrate curricula. *Decision Sciences Journal of Innovative Education*, 13(2), 151–176.

Sroufe, R., Sivasubramaniam, N., Ramos, D., and Saiia, D. (2015). Aligning the PRME: How study abroad nurtures responsible leadership. *Journal of Management Education*, 39(2), 244–275.

Sroufe, R., Stevenson, C. E., and Eckenrode, B. A. (2019). *The power of existing buildings: Save money, improve health, and reduce environmental impacts*. Washington DC: Island Press.

Starik, M. (2015). The downside of (and antidote to) unhelpful stealth sustainability. *Organization & Environment*, 28(4), 351–354. doi: https://doi.org/10.1177/ 1086026615623589

Strauss, A., and Corbin, J. M. (1997). *Grounded theory in practice*. New York, NY: SAGE Publications.

United Nations. (2015). Transforming our world: The 2030 agenda for sustainable development (A/RES/70/1), accessed June 7, 2021 at https://sustainabledevelopment .un.org/index.php?page=view&type=111&nr=8496&menu=35

United States Department of Agriculture. (2020). Economic Research Services, Key Statistics & Graphics, Food Security in the U.S., accessed June 7, 2021 at https:// www.ers.usda.gov/topics/food-nutrition-assistance/food-security-in-the-us/key -statistics-graphics.aspx

US Energy Information Administration. (2020). *Commercial buildings energy consumption survey (CBECS), 2018 preliminary results*, November, Office of Energy Statistics, U.S. Department of Energy, Washington, DC 20585.

World Green Building Council. (2020). Green building: Improving the lives of billions by helping to achieve the U.N. Sustainable Development Goals, accessed August 1, 2020 at https://www.worldgbc.org/news-media/green-building-improving-lives -billions-helping-achieve-un-sustainable-development-goals

4. Teaching complex adaptive systems through multiple spheres of influence

Bernadette Roche

SOCIAL–ECOLOGICAL SYSTEMS AND THE UN SUSTAINABLE DEVELOPMENT GOALS

A love of nature has pervaded my daily activities for as long as I can remember. An important part of my development was coursework and research at the University of Virginia's Mountain Lake Biological Station. At age 19, I found my tribe there, with dinner conversations about spiders, oaks, and owls, and complete immersion into the natural world of the southern Appalachians. Nine summers there propelled me to preserve and support biodiversity throughout my life.

The global environmental system has shifted from nature-dominated to human-dominated (Rockström et al., 2009); those working at the social–ecological interface must address the accelerating impacts of humans on the entire biosphere in this new epoch, the Anthropocene (Cruzan, 2006). Complexity of social–ecological systems requires a shift from reductionist thinking emphasizing individual actions, to a systems mindset emphasizing interconnectedness, emergence, and nonlinearities. I currently take many sustainable actions suggested by Kanashiro et al. (2020), such as reducing fossil fuel use, eating a more plant-based diet, and reducing water consumption, as well as incorporating sustainability into courses and developing the Environmental and Sustainability Studies Minor at Loyola University Maryland (LUM). The framework of the UN Sustainable Development Goals (SDGs) (United Nations, 2015) makes a strong call for greater involvement of ecologists, among others. Two goals are explicitly about ecological integrity (#14 life below water, #15 life on land) and others relate to social–ecological systems: #6 clean water and sanitation; #7 affordable clean energy; #9 industry, innovation, and infrastructure; #11 sustainable cities and communities; #12 responsible consumption and production; #13 climate action. As I share with my students, the remaining goals resonate with me as a person of faith: #1 no poverty; #2 zero hunger; #3 good health and well-being; #4 quality education; #5 gender equality; #8 decent work and economic growth; #10 reduced inequalities; #16 peace, justice and strong institutions; #17 partnerships for the goals. Though they have much lower per-capita carbon footprints (Hubacek et al., 2017), the poor are both disproportionately impacted by climate change and have the least capacity to respond to the impacts (Levy & Patz, 2015). Thus, we are compelled to act in solidarity with those suffering the greatest impacts.

TEACHING COMPLEXITY TO MEET UN SDGS

What makes teaching sustainability so challenging, beyond the suggestions for lifestyle choices to live more lightly on the land? Social–ecological systems, where social issues and ecological issues overlap, are by their very nature complex adaptive systems (Prejser et al., 2018). Treating them as two separate fields rather than one integrated whole where "humans and nature are deeply intertwined" leads to the problem of missing important aspects of the dynamics of the system (Hertz et al., 2020). The complex adaptive systems inherent in the SDGs cannot be addressed with reductionistic thinking and have properties that are not immediately inviting to undergraduate students: nonlinearities, heterogeneity, far-from-equilibrium dynamics, interactions and processes, and contingencies (Rogers et al., 2013). Reductionism relies on deconstructing systems into separable and independent units, with a linear cause-and-effect structure. Such knowable structure and behavior enable scientists to provide stakeholders with mappable pathways backed by undisputable facts. While more tractable for undergraduates, these structures do not reflect reality. Reductionist thinking distorts our view of reality, but very few people truly embrace complexity thinking. Rogers et al. (2013) suggest some "habits of mind" that facilitate complexity thinking to enhance approaches to addressing SDGs. The frames of mind are grouped into those associated with an open mind/openness of behavior (e.g., hold strong opinions lightly, accept surprise, be open to both/and options, expect and accept ambiguity and paradox); those that promote situational awareness (e.g., consideration of interactions and not just components, note contingencies, scale, temporal/spatial context and history, iterative individual and collective reflection); and a balance between decisiveness and restraint (e.g., courageous thinking, seizing the moment, embracing mistakes as part of the learning process, honoring the group dynamic, and considering all options on the table for longer than is comfortable).

In this "sound-byte" and "tweet-byte" society, adopting these habits of mind requires a slowing down, and a recognition of the multiplicity of legitimate perspectives involved in complex adaptive systems with STEEP (social, tech-nical, economic, environment, and political) attributes (Rogers et al., 2013).

Asking academics to "hold their strong opinions lightly" (Rogers et al., 2013) is perhaps the most challenging aspect of full adoption of these habits of mind. Why not just simply teach the content of the course as facts from an expertise stance? In the science classroom, students are taught an array of facts from textbooks that are out of date before they are even printed. Faculty can transmit these facts, but many recognize that it is better to teach the skills of knowledge acquisition, teaching students how to research and act on their ideas. Even this approach, without the benefits of open and reflective thinking, can backfire when it comes to creating critical thinkers for solving the problems of the future, because those with the highest scientific literacy skillfully find studies to back their own ideologies (Amel et al., 2017), but without concomitant discernment. The habits of mind that promote complexity thinking increase dialogue and inquiry, and certainly promote the type of problem solving that is required for the complex transdisciplinary issues of today (Schoon & Van der Leeuw, 2015).

Many faculty have adopted a teacher–scholar identity, especially those at primarily undergraduate institutions. The conception of faculty member as teacher–scholar stems from Boyer's (1990) four functional categories of scholarship: discovery, community engagement, application to societal needs, and teaching. The teacher–scholar in Boyer's (1990) model incorporates teaching and discovery of knowledge with dissemination and partnerships beyond the traditional avenues, including community engagement and societal applications. Teacher–scholars that include undergraduate students in research benefit both themselves and the students (Gardner et al., 2010). Students are more likely to engage in research when faculty place a high value on it (Kuh et al., 2007). There is every reason to believe that this would apply to community engagement as well. Because we live in a time where we are exceeding Earth's safe operating space (Rockström et al., 2009), we must strive for teacher–scholar–practitioners with a moral commitment to social justice (Amonett, 2014). The teacher–practitioner divide (McNatt et al., 2010) must be dissolved, so that students see an authentic reflection of what is being taught (conceptual matters) in what is being done (reality). Faculty teaching about sustainability must "walk the walk" and engage their students to do the same (Kanashiro et al., 2020). In this rapidly changing landscape, with accelerating impacts (Steffen et al., 2015), faculty are often learning along with the students, and as such, faculty can embrace hooks' (2013) philosophy placing the teacher–scholar as a continual co-learner in community with their students. This teaching style has the added benefit of promoting equity and inclusion for a diverse student body (SDG #10).

I have certainly become a better teacher–scholar–practitioner as a result of shifting my sustainability actions into full gear in my personal life and academic life. After going to the Association of American Colleges and

Universities Conference (AAC&U) STEM education conference (AAC&U, 2015), I was inspired by the idea of teaching ecology through the conceptual framework of threshold concepts, especially to address conceptually challenging material in social–ecological systems. Ecology courses are replete with threshold concepts (Hill, 2010; Loertscher et al., 2014), defined as transformative, troublesome, irreversible, integrative, and difficult to grasp. Key to teaching threshold concepts is insight that learners stand on one side of a "portal" (Meyer & Land, 2006) with incomplete understanding, and faculty stand on the other side, with full understanding but lacking the memory of a time when they could not grasp the particular threshold concept. When faculty recognize the irreversible nature of a particular threshold concept (i.e., that something which seems obvious to them is not obvious to the learner), they can help guide students through the "portal". I created clicker questions (using online polling software, Poll Everywhere) to help students see where their thinking was first, erroneous, and second, shifting. This approach alleviated some of the discomfort students feel when confronted with their lack of mastery of complex issues, because they could remain anonymous and learn from each other's mistakes. In other words, it gave freedom to make mistakes, attention to how to correct them, and newfound mastery of concepts, through public discussion of the anonymous clicker answers. Students asked for more clicker questions when I asked for their opinion, and so I have made a habit of including them in just about every class I teach.

A new avenue of teaching that I am exploring, by taking part in the Bonner Foundation's "Social Action Couse Development" webinar series, is to incorporate social action campaigns related to sustainability in one or more of my courses. This approach to "walking the walk" resonates with me because it recognizes students as co-learners and provides them with agency to make meaningful and beneficial changes related to the SDGs. They become doers rather than learners, which seems to be the point of all education.

EXPANDING SPHERES OF INFLUENCE FOR IMPACTFUL APPROACHES TO SUSTAINABILITY

A conceptual model for inspiring impactful changes is the hierarchical spheres of influence rippling out from personal, to social network, organizational, public, and cultural spheres (Amel et al., 2017). The systemic changes required to meet the SDGs depend on shifts in societal norms, made via the cultural sphere of influence. While I have incorporated sustainable practices with a variety of personal lifestyle changes (e.g., composting, eating a more plant-based diet, driving a hybrid car, installing solar panels), the most impactful change my husband and I have made, and shared with students, is a true commitment to promoting biodiversity on our property through a systemic change in how we manage our landscape: removing nonnatives, planting native plants, and refraining from using pesticides or herbicides. By our supporting SDG #15, the entire ecosystem in our yard has changed: no more pest outbreaks, and many beneficial insects such as pollinators, other invertebrates, and birds. As a result of this effort, my husband is now a member of our town sustainability committee, sharing our approach. There are many areas where I could improve in my personal sustainability efforts. I enjoy scuba diving, taking long flights about once per year. I have thought about doing offsets for air travel, but I have not yet consistently done that. Though I was once a vegetarian, I eventually "fell off the carrot" and switched back to eating meat. I have pushed the diets of my husband and myself to a more local and plant-based diet, but we could do better. These confessions cause me a bit of shame, in that my ecological footprint is larger than it should be. However, though I always wish to do better within my own personal sphere of influence, I preserve some of my energy to be more impactful through larger spheres of influence.

It has been straightforward as a professor who teaches sustainability to move out from the private sphere to the social network sphere, where sharing expertise with students is part of the job. Asking students to start with their own personal sphere of influence is empowering to them, and they begin to see how impactful their sustainable choices can be. With some encouragement, they often move out towards the next ripple, of social network, influencing other students via clubs such as the Environmental Action Club at LUM. Some students will even find their way into the organizational ripple sharing their ideas with administration and demanding procedural changes on campus sustainability. Likewise, professional responsibilities for professors could, for example, involve a move into the organizational ripple by being part of a sustainability committee (Denis & Gannon, 2014). I was chair of our LUM sustainability committee formed in January 2016 in support of the President's Carbon Commitment, signed by our university president in 2015 and followed by a Climate Action Plan in 2018, and I have encouraged other faculty to take on this role. Partnering with others at LUM with similar environmental justice concerns, I have hosted films for the Baltimore Environmental Film Festival at LUM and have invited speakers for the Environmental and Sustainability Studies Minor (Brian Czeck on Environmental Economics, Rear Admiral David Titley on Climate Change, Doug Tallamy on Bringing Nature Home). Both faculty and students should recognize that although individual actions and influences on one's social network and organization are important and empowering, they are not sufficiently impactful to address the greatest challenges outlined by the SDGs.

I have incorporated community assets into my teaching and research in a variety of ways. In 2013 my Plant Ecology lab course included a service-learning project mapping the plants in a pollinator garden for Audubon of Patterson Park. I participated in Cornell's Urban Environmental Education Online Professional Development Community, which culminated in two collaborative book chapters on schoolyard habitats for birds (Roche & Agnello, 2015) and promoting environmental action (Roche et al., 2015). I recently launched a new large-scale, long-term research project to attract biology majors and non-majors from the Environmental and Sustainability Studies Minor to do biodiversity studies. This research is part of the Ecological Research as Education (EREN) Permanent Forest Plot Project (PFPP) involving a collaboration with Lake Roland, a Baltimore city/county park. To deal with the complex adaptive system of urban forest dynamics, I am collaborating with a colleague in New Jersey to apply Structural Equation Modeling (SEM) to better understand the forest system over space and time (Morrison, 2017). SEM is a systems approach for distinguishing between alternative hypotheses of how the system works, addressing direct and indirect interactions not only among measured variables, but also among latent variables representing unmeasurable concepts (e.g., health and well-being). Teaching students this powerful technique emphasizes the habits of mind for complexity thinking, demonstrating to them how to address complex systems with a set of multiple plausible hypotheses rather than one best hypothesis, and it may entice them to take action through future research projects of their own.

The greatest impacts an individual can have via their multiple rippling spheres of influence are when their actions influence broader systems, moving beyond their own social network and organizations, and out to public and cultural arenas (Amel et al., 2017). Early in 2020, I began to give sustainability talks, presenting at my local library and at a regional high school. In order to push my sphere further out in the public and cultural realms, I have connected my professional expertise to leadership to my religious faith, to focus local faith communities on the moral obligation for environmental stewardship. I dedicate much of my spare time to my community as co-lead of both GreenGrace (the Environmental Ministry of the Episcopal Diocese of Maryland) and MACS (the Multi-faith Alliance of Climate Stewards in Frederick, MD). Community-based social marketing (McKenzie-Mohr, 2020) has the ability to propel people into action, as they begin to see that their own behavior goes against the norm. Feeling supported, that individual contributions matter, and that solutions are possible (Amel et al., 2017) helps propel activism and behavior changes to combat climate change. "Hope in the midst of tragedy" (Pihkala, 2018), focusing on constructive or realistic hope that humans can change course to create a more sustainable future (Marlon et al., 2019), is a powerful motivating factor for action. While fatalistic doubt may lead to despair and inaction, constructive doubt can play a role in demonstrating the urgency of a rapid and large response before it is too late and humanity is doomed (Marlon et al., 2019).

GreenGrace promoted tree plantings with the Maryland/Delaware Lutheran Synod, and educated the diocese with video productions, on plastic pollution and on the importance of planting trees. I published a letter to the editor with my GreenGrace co-lead Dick Williams in the *Baltimore Sun* supporting Maryland forest conservation legislation (Williams & Roche, 2019). In partnering with the Maryland Environmental Human Rights (MDEHR) campaign, Dick and I became founding members of the Faith Communities Working Group for the MDEHR, helping with the implementation of a faith-based Environmental Justice (EJ) webinar series to build solidarity for environmental human rights issues in our state. Participants were given a background in the history of EJ, information on environmental health disparities, tools for advocacy, action, and allyship, and were able to build strong networks for future action. The ultimate goal of this work is to pass an Environmental Human Rights Amendment in the state of Maryland, empowering individuals to take legislative action against entities that have infringed upon their environmental human rights (MDEHR, 2020). The MDEHR strives to make the type of system-wide change needed to address the SDGs related to social–ecological systems, such as sustainable cities and communities, as well as the SDGs of health and well-being and peace, justice, and strong institutions. As a result of my work there, several LUM students are currently doing internships with the campaign.

Other system-wide changes addressing environmental rights have been attained through my work with MACS, focusing on several campaigns aimed at the Frederick County Council. The Council has responded by passing the Climate Emergency Resolution and two ordinances: a Zoning Ordinance (protects ecologically significant forests from development) and a Forest Restoration Ordinance (ensures equal acreage of replacement forests for every acre developed). Through an outreach program in collaboration with Interfaith, Power & Light this summer, we presented three Eventide events presented as half-hour Zoom events with prayer, song, and calls to action for faith communities throughout Frederick County, MD. Action items included writing to council members on the two ordinances and participating in a community solar outreach. Working with faith communities is a way to promote cultural shifts, to make it the norm for these communities to focus on environmental justice, which is already consistent with the tenets of all major religious organizations. The United Nations recognizes that faith-based organizations are critical to aid in the social mobilization required to move towards meeting the UN SDGs (Karam, 2014), and this can be achieved through top-down approaches like meeting with religious leadership, or through bottom-up approaches like meeting with local multi-faith community groups.

A key theme to moving beyond organizational spheres into public and cultural spheres within supportive organizations is the motivating force of collective action. One does not have to have a charismatic personality to bring about societal changes; rather, becoming an integral part of an organization composed of many people working towards a common cause can result in large changes. Leadership is required, but the collective act of large numbers of people can promote swift and sweeping changes. Why do cultural norms have to shift? Because although a record number of Americans are alarmed or concerned about climate change (Goldberg et al., 2020), far fewer feel that the actions required to combat it are worth paying for (Reuters/Ipsos, 2019).

LINKING BIODIVERSITY AND HUMAN WELL-BEING

Teaching about biodiversity is my greatest joy, but also a source of great sadness as I realize that many of my students are so disconnected from nature that they seem to have little interest in preserving it. It is important to me that students grasp how important biodiversity is to sustaining life on earth. Heterogeneity is the key to the adaptiveness of complex adaptive systems such as our biosphere (Levin, 1998), and biodiversity is the source of heterogeneity in the ecosystem. Too much biodiversity loss, and the biosphere potentially moves past tipping points into spaces less conducive to human life (Rockström et al., 2009). Engaging students in nature can be challenging: students ask, "Are there going to be bugs?" when discussing our field trips to natural areas near LUM, and I always respond, "I sure hope so!" I have begun to market the value of nature to my students, demonstrating the known benefits of nature for human health and well-being (Louv, 2016; White et al., 2019), linked to SDG #3. I assign students to walk on a nature trail, and then write a reflection on how it impacted their mental state. Beyond that, I want students to really appreciate the beauty and relevance of the biodiversity around them. I recently ran a BioBlitz using iNaturalist (2020) with LUM students in the Ecology, Evolution, and Diversity lab course, where students observed or identified 10 distinct naturally occurring taxa, and student feedback was positive.

In conclusion, sustainability is not separable from my identity as a teacher, scholar, and activist. It has not always been this way, and I have evolved to respond to the needs that I see in the world where they meet my expertise as an ecologist. Since it is now clear that humans are altering the planet in ways that make life less hospitable for themselves, we all have to heed the call to action to reverse course.

REFERENCES

AAC&U. (2015). Crossing Boundaries: Transforming STEM Education, accessed on January 18, 2021 at https://www.aacu.org/meetings/stem/15

Amel, E., Mannning, C., Scott, B., & Koger, S. (2017). Beyond the roots of human inaction: Fostering collective effort towards ecosystem conservation. *Science, 356*(6335), 275–279. https://doi.org/10.1126/science.aal1931

Amonett, C. Y. (2014). Moral commitment: Scholar–practitioners making choices with strength of purpose. In P. M. Jenlink (Ed.), *Educational leadership and moral literacy: The dispositional aims of moral leaders* (pp. 55–68). Lanham, MD: Rowman & Littlefield.

Boyer, E. L. (1990). *Scholarship reconsidered: Priorities of the professorate.* Carnegie Foundation for the Advancement of Teaching.

Cruzan, P. (2006). The "Anthropocene". In E. Ehlers & T. Krafft (Eds.), *Earth system science in the Anthropocene* (pp. 13–18). New York, NY: Springer. https://doi.org/10.1007/3-540-26590-2_3

Denis, C., & Gannon, A. (2014). Walking the eco-talk movement: Higher education institutions as sustainability incubators. *Organization & Environment, 27*(1), 16–24. https://doi.org/10.1177%2F1086026614521629

Gardner, J. C., McGowan, C. B., & Moeller, S. E. (2010). Applying the teacher–scholar model in the school of business. *American Journal of Business Education, 3*(6), 85–89. https://doi.org/10.19030/ajbe.v3i6.446

Goldberg, M., Gustafson, A., Rosenthal, S., Kotcher, J., Maibach, E., & Leiserowitz, A. (2020). For the first time, the Alarmed are now the largest of Global Warming's Six Americas, Yale Program on Climate Change Communication, accessed on January 18, 2021 at https://climatecommunication.yale.edu/publications/for-the-first-time-the-alarmed-are-now-the-largest-of-global-warmings-six-americas/

Hertz, T., Garcia, M. M., & Schlüter, M. (2020). From nouns to verbs: How process ontologies enhance our understanding of social–ecological systems understood as complex adaptive systems. *People and Nature, 2*(2), 328–338. https://doi.org/10.1002/pan3.10079

Hill, S. (2010). Troublesome knowledge: Why don't they understand? *Health Information and Libraries Journal, 27*(1), 80–83. https://doi.org/10.1111/j.1471-1842.2010.00880.x

hooks, b. (2013). *Teaching community: A pedagogy of hope.* Abingdon: Routledge. https://doi.org/10.4324/9780203957769

Hubacek, K., Baiocchi, G., Feng, K., Castillo, R. M., Sun, L., & Xue, J. (2017). Global carbon inequality. *Energy, Ecology, and Environment, 2*(6), 361–369. https://doi.org/10.1007/s40974-017-0072-9

iNaturalist. (2020). Home page, accessed on January 8, 2021 at https://www.inaturalist.org/

Kanashiro, P., Rands, G., & Starik, M. (2020). Walking the sustainability talk: If not us, who? If not now, when? *Journal of Management Education, 44*(6), 822–851. https://doi.org/10.1177%2F1052562920937423

Karam, A. (Ed.). (2014). Religion and development post-2015. United Nations, accessed on January 18, 2021 at https://www.unfpa.org/sites/default/files/pub-pdf/DONOR-UN-FBO%20May%202014.pdf

Kuh, G. D., Chen, D., & Laird, T. F. N. (2007). Why teacher–scholars matter: Some insights from FSSE and NSSE. *Liberal Education, 93*(4), accessed January 18, 2021 at https://www.aacu.org/publications-research/periodicals/why-teacher-scholars -matter-some-insights-fsse-and-nsse

Levin, S. A. (1998). Ecosystems and the biosphere as complex adaptive systems. *Ecosystems, 1*(5), 431–436. https://doi.org/10.1007/s100219900037

Levy, B. S., & Patz, J. A. (2015). Climate change, human rights, and social justice. *Annals of Global Health, 81*(3), 310–322. http://doi.org/10.1016/j.aogh.2015.08.008

Loertscher, J., Green, D., Lewis, J. E., Lin, S., & Minderhout, V. (2014). Identification of threshold concepts for biochemistry. *CBE–Life Sciences Education, 13*(3), 516–528. https://doi.org/10.1187/cbe.14-04-0066

Louv, R. (2016). *Vitamin N: The essential guide to a nature-rich life.* Chapel Hill, NC: Algonquin Books.

Marlon, J. R., Bloodhart, B., Ballew, M. T., Rolfe-Redding, J., Roser-Renouf, C., Leiserowitz, A., & Maibach, E. (2019). How hope and doubt affect climate change mobilization. *Frontiers in Communication, 4,* 20. https://doi.org/10.3389/fcomm .2019.00020.

Maryland Environmental Human Rights. (2020). Home page, accessed on January 18, 2021 at mdehr.com

McKenzie-Mohr, D. (2020). Fostering sustainable behavior through community-based social marketing. *American Psychologist, 55*(5), 531–537. https://doi.org/10.1037/ 0003-066X.55.5.531

McNatt, D. B., Glassman, M., & Glassman, A. (2010). The great academic–practitioner divide: A tale of two paradigms. *Global Education Journal, 23*(2), 6–22.

Meyer, J., & Land, R. (Eds.) (2006). *Overcoming barriers to student understanding: Threshold concepts and troublesome knowledge.* London: Routledge. https://doi .org/10.4324/9780203966273

Morrison, J. (2017). Effects of white-tailed deer and invasive plants on the herb layer of suburban forests. *AOB Plants, 9*(6), plx058. https://doi.org/10.1093/aobpla/plx058

Pihkala, P. (2018). Eco-anxiety, tragedy, and hope: Psychological and spiritual dimensions of climate change. *Zygon, 53*(2), 545–569. https://doi.org/10.1111/zygo.12407

Prejser, R., Biggs, R., De Vos, A., & Folke, C. (2018). Social–ecological systems as complex adaptive systems: Organizing principles for advancing research methods and approaches. *Ecology and Society, 23*(4), 46. https://doi-org.proxy-ln.researchport .umd.edu/10.5751/ES-10558-230446

Reuters/Ipsos. (2019). Americans' attitudes on climate action, accessed on January 18, 2021 at https://graphics.reuters.com/USA-ELECTION-CLIMATECHANGE/ 0100B03104Z/index.html

Roche, B., & Agnello, J. (2015). Birds and schoolyard habitats. In A. Russ (Ed.), *Urban environmental education* (pp. 100–103). Cornell University Civic Ecology Lab, NAAEE and EECapacity.

Roche, B., Brown, J., & Shmulenson, I. (2015). Promoting environmental action competence through urban environmental stewardship. In A. Russ (Ed.), *Urban environmental education* (pp. 104–107). Cornell University Civic Ecology Lab, NAAEE and EECapacity.

Rockström, J., Steffen, W., Noone, K., Persson, Å., Chapin III, F. S., Lambin, E. F., Lenton, T. M., Scheffer, M., Folke, C., Schellnhuber, H. J., Nykvist, B., de Wit, C. A., Hughes, T., van der Leeuw, S., Rodhe, H., Sörlin, S., Snyder, P. K., Costanza, R., … Foley, J. A. (2009). Planetary boundaries: Exploring the safe operating space for humanity. *Ecology and Society, 14*(2), 32. https://doi.org/10.5751/es-03180-140232

Rogers, K. H., Luton, R., Biggs, H., Biggs, R. (Oonsie), Blignaut, S., Choles, A. G., Palmer, C. G., & Tangwe, P. (2013). Fostering complexity thinking in action research for change in social–ecological systems. *Ecology and Society*, *18*(2), 31. https://doi.org/10.5751/es-05330-180231

Schoon, M., & Van der Leeuw, S. (2015). The shift toward social–ecological systems perspectives: Insights into the human–nature relationship. *Natures Sciences Sociétés*, *23*(2), 166–174. https://doi.org/10.1051/nss/2015034

Steffen, W., Broadgate, W., Deutsch, L., Gaffney, O., & Ludwig, C. (2015). The trajectory of the Anthropocene: The great acceleration. *Anthropocene Review*, *2*(1), 81–98. https://doi.org/10.1177/2053019614564785

United Nations. (2015). Transforming our world: The 2030 agenda for sustainable development (A/RES/70), accessed on June 18, 2021 at https://sustainabledevelopment.un.org/index.php?page=view&type=111&nr=8496&menu=35

White, M. P., Alcock, I., Grellier, J., Wheeler, B. W., Hartig, T., Warber, S. L., Bone, A., Depeldge, M. H., & Fleming, L. E. (2019). Spending at least 120 minutes a week in nature is associated with good health and wellbeing. *Scientific Reports*, *9*(1). https://doi.org/10.1038/s41598-019-44097-3

Williams, D., & Roche, B. (2019). Time to close loopholes on Md.'s forest conservation law. *Baltimore Sun*, February 21, accessed on January 18, 2021 at https://www.baltimoresun.com/opinion/op-ed/bs-ed-op-0222-forest-conservation-20190221-story.html

5. If everything is connected, where do you begin?

Jimmy Y. Jia and Rick Dickinson

As educators of sustainability, we find it easy to point out the non-linear complexities of human impact on the climate, yet these complexities are abstract in daily life. Our socioeconomic systems are stressing our planet's environmental and climate systems to a tipping point. Human systems and impacts have proliferated dramatically in the last century, with new settlements and increased economic activities. Meanwhile, environmental destruction, such as deforestation and the draining of wetlands, has decreased the overall resiliency of the environmental systems. This is manifesting itself in numerous ways, including dramatic Arctic warming (Thoman et al., 2020) and an increase in the number of billion-dollar weather-related disasters experienced across the world (MunichRE, 2017).

Yet, *which* personal daily activities contribute to the global climate-related disaster is much more difficult to pinpoint. What individuals do on a day-to-day basis is commute to work, buy groceries, and spend time with family and friends. Daily habits and choices may seem negligible to global patterns, but they are a root cause of many environmental challenges. At the planetary level, scientists observe the aggregated effects of these individual actions, including ocean acidification, sea level rise, and the loss of biodiversity.

Sustainability professionals, whether educators or practitioners, face several quandaries. Any sustainability initiative may improve a local or regional system, but the planetary risks could remain the same. Conversely, a logical solution to address planetary risk may make a local system less efficient or worse off (Öberg et al., 2012). Solutions may benefit certain stakeholders while disadvantaging others. At best, these system interactions will confound students who are grappling with the issue for the first time (Savageau, 2013). At worst, our leaders will be stymied by the complexities of collective action and fail to take action (Finke et al., 2016).

In this chapter, the authors relate a story of how one author developed a curriculum to teach MBA students how to mitigate the climate risks of corporate decisions. Much like how Theseus in Greek mythology used a red thread to navigate a labyrinth to slay the Minotaur, the curriculum traces the red threads of sustainability decisions through the labyrinth of the socio-environmental-economic system. The key lesson of the curriculum is that within complex systems, individuals can only affect the subsystems they touch. However, those subsystems interact with other subsystems, and so on, extending the individual's impact to systems they don't directly influence.

The other author, a former student, recognized that the decision wholly within his control was his personal consumption choices. Based on the lessons of the class, he developed and adopted his Rubric for Sustainable Personal Consumption that met the goals of two UN Sustainability Development Goals (UN SDGs): *UN SDG 10: Reduce Inequalities Within and Among Countries* and *UN SDG 12: Responsible Consumption and Production.*

When the first author first learned of the new rubric, he added the rubric to the MBA curriculum as an example of how corporate climate risk strategies can be modified for an individual's use. It became a lesson in how everyday decisions could create new reinforcing feedback loops at the individual level to change the macroscopic patterns of the whole system (Lucas et al., 2018). Furthermore, the collaboration between the authors to write this chapter has culminated in a more refined rubric as well as both authors becoming more aware of their sustainability practices. As a result, both have more intentionally embedded these practices into their teachings, professional work, and personal habits.

TRAVERSING THE SUSTAINABILITY LABYRINTH

The interconnectedness of our socio-environmental-economic system makes them a challenge to fully understand. If the climate system wasn't complex enough on its own, there's been a number of studies examining the nexuses of energy–poverty–climate (Casillas & Kammen, 2010), water–food–energy–climate (Waughray, 2011), and energy–transportation–air quality–climate change and health (Erickson & Jennings, 2017), to name a few. To a student, these overlapping systems can be confusing.

To help students untangle the nests of subsystems within sustainability, one of the authors developed a four-step process that he named *Tracing the Red Threads*:

1. *Define a balanced system.* The socio-environmental-economic system is the labyrinth being explored.
2. *Pick one primary subsystem and analyze it deeply.* Theseus tied his red thread to the doorpost of the labyrinth, giving him a reference point. This step anchors the analysis to a subsystem as a reference point.

3. *Examine the effects on adjacent subsystems.* From the reference point of the primary subsystem, examine the adjacent secondary subsystems, tertiary subsystems, and so on.
4. *Move the analysis to a new primary subsystem and repeat steps 2–4.* "Tie" the red thread to a different primary subsystem and explore the system again.

Step 1: Define a Balanced System

The Energy Literacy Matrix was created to frame the socio-environmental-economic system being studied. At its core, the matrix combines a technical framework that describes how energy flows and a societal framework that describes how societies manage limited resources.

As a technical issue, energy is in balance through the supply chain. The source of energy is the fuel used, such as coal or sunlight. The sink is how the energy gets consumed, such as residential, industrial, or commercial end uses. Rarely is the source of fuel geographically next to the end user of the sink. Thus, one also needs to examine transportation or electric transmission infrastructure and their accompanying financial mechanisms to move energy from the source to the sink. For example, the canal, rail, and road networks of the UK were developed in the mid-1800s as a way to transport coal from the mines to city centers (Jevons, 1866), and in North America, the centralized electric grid was first built to move power from Niagara Falls to Buffalo in 1890 (Jia & Crabtree, 2015).

Decarbonizing the energy system therefore has multiple intervention points. For example, Washington State published a report in 2016 analyzing deep decarbonization pathways. Scenarios analyzed included tapping into onshore wind and solar panels (interventions at the source), demand-side equipment sales (interventions at the sink), and electrifying transportation (interventions in the transport of goods) (Haley et al., 2016).

As a social issue, energy is a Wicked Problem. Wicked Problems are those in which the solution changes based on the problem framing (Rittel & Webber, 1973). Furthermore, the problem frame also changes once a solution is applied, continuing to evolve the problem. As an example, in the early 1900s, horses were the main mode of transportation in New York City. The 100,000 horses in the city produced over 2.5 million pounds of manure per day (Burrows & Wallace, 2000). One solution to this pollution problem was the horseless carriage, or automobile. Today, the transportation pollution problem still exists, except it has shifted from horse manure to emissions caused by the automobiles themselves (Watson et al., 1988).

Wicked Problems require a Clumsy Solution approach (Verweij & Thompson, 2006). Rather than elegant, single-mode solutions, Clumsy

Solutions suggest that there needs to be a balance of hierarchical, competitive, and egalitarian approaches. Hierarchical approaches use processes and rules to create the pathway for an outcome. An example is when a public utility commission manages the process for public commenting before determining a new utility rate structure. Competitive approaches are those that find the "best" solution using competitive mechanisms. Businesses that compete for market share or entrepreneurs seeking to disrupt a sector with a better tool exemplify competitive approaches. Egalitarian approaches are those that advocate for socially beneficial or ideal solutions. Examples may include non-profits that promote renewable energy outcomes and activists advocating for the protection of wetlands.

Clumsy Solutions appear inelegant because one may need participation of all three approaches to make progress (Grint, 2010). For example, take an environmental group (egalitarian) that advocates for clean energy. The group may be more effective if they engaged with a public utility commission's process (hierarchical) to approve a financial incentive that helps businesses sell more solar panels (competitive).

The Energy Literacy Matrix is the result of merging the energy supply chain with the Clumsy Solution. Each cell of the matrix is a subsystem that can be analyzed individually, while the entire matrix demonstrates the interconnected whole. Table 5.1 depicts a selection of the MBA classes taught using this framework. The topics seem disparate – from international nuclear treaties to advanced manufacturing – but the students are able to see how each energy issue is related to all others. The list is not exhaustive by any means and many other classes could be taught in any of the cells. For example, other governmental approaches to generation issues could include policies for fuel subsidies, renewable energy mandates, and Paris Treaty alignment. Other business solutions to end uses could include internet of things (IoT), demand response, and sustainable product development. Instructors are encouraged to map their own curriculum to the matrix to diagnose the broadness of coverage in their lessons. For general sustainability classes, the matrix can help ensure that no single cell is overly emphasized with too many lessons.

Step 2: Pick One Primary Subsystem and Analyze It Deeply

Each class is anchored in one of the cells of the Energy Literacy Matrix. Students study in detail one sector, such as electricity regulations, water extraction technologies, or carbon pricing policy. This helps students understand the strengths, weaknesses, and constraints of a specific subsystem. For instance, in the lecture on how water districts serve their community, the discussion starts with the internal functionality of the water district – the role of the commissioners, executives, engineers, and so on. Very quickly,

Table 5.1   *Example MBA classes taught using the Energy Literacy*
                Matrix

	Production / generation	Transportation / transmission	Consumption / end uses
Hierarchical (i.e., government)	• International nuclear security • Fuel taxes & incentives	• Pipeline siting • Public transit policy	• Public utility commissions • Distributed generation
Competitive (i.e., business)	• Financial structures of renewable PPAs • Angel investing of renewable resources	• District heating • Wholesale electricity markets	• Advanced manufacturing • Energy efficiency
Egalitarian (i.e., non-profit)	• Developing world • Carbon advocacy	• Electric vehicle advocacy	• Water-as-a-right • Zero waste

students would see how water from snow melt in the mountains is not only Washington's true regional water supply, the same water is also needed for ecological preservation of salmon habitat, cultural needs of regional tribes, and a provider of the region's clean hydropower (BPA, 2020).

Step 3: Examine the Effects on Adjacent Subsystems

Next, students examine the dynamics of how the primary subsystem affects adjacent subsystems. For example, a public utility commissioner once acknowledged in class that their mandate for authorizing rate increases based on prudent capital investments of electricity infrastructure was stifling innovation. He suggested that perhaps the utility regulatory model should learn from the regulation of transportation, where many innovations in electric and autonomous vehicles have been initiated. Several months later, when a state's transportation official was the guest, she disagreed with the suggestion. One problem, she noted, with the transportation regulatory model is that the infrastructure funding depends on legislative action rather than business needs. Thus, necessary improvements to transportation services often become politicized.

Step 4: Move the Analysis to Another Primary Subsystem

To ensure broad coverage of topics, each class is anchored in a different cell of the Energy Literacy Matrix. Students would slowly create a map of how the issues were related. Even if the same topic were taught twice, such as water, it would be taught using two different lenses. For instance, the sustainability issues faced by a city's Public Works Department to provide fresh water are very different from the sustainability issues to dispose of stormwater.

The class has generated several resources available to other educators, namely two books (Jia, 2020b; Jia & Crabtree, 2015), a podcast series (Jia, 2020a), and a GitHub repository of open-sourced lecture notes (Jia, 2018).

KEY TEACHING POINTS FROM THE ENERGY LITERACY MATRIX

Students are asked to analyze and recommend *decision priorities* in order to resolve tensions between different cells of the Energy Literacy Matrix. For example, electric utilities are managed by public utility commissions (cell: Hierarchical / Generation). Roads are managed by state departments of transportation (cell: Hierarchical / Transport). The telecommunications sector, which consumes electricity, is regulated by the US Federal Communications Commission (FCC) (cell: Hierarchical / Consumption). Under which regulatory body should an autonomous electric vehicle fall, which requires the support of the electrical, transportation, and telecommunication subsystems?

Students study how *risk minimization strategies* tend to also minimize climate risks. For example, the traditional insurance company's approach to mitigating risks is to diversify the portfolio. By having non-coupled risks, no single calamity will bankrupt the company. However, reinsurers, or those who provide insurance to insurance companies, are exposed to risks in nearly every sector, geography, and market. They cannot diversify away from global climate-related risks. Instead, the sector must reduce their global impact to minimizing global risks. Students study a modified version of the United Nations Environment Programme's recommended framework for minimizing the exposure to systemic risks (UNEP, 2009):

1. *Avoid* taking actions that generate the risk.
2. If one needs to take action, *minimize* the activities that generate the risk.
3. Once one takes action, *manage* the risks that are generated.

PERSONAL SUSTAINABILITY LESSONS

One of the authors, a student who studied energy systems with the above methodology, used the key teaching points to develop his Rubric for Sustainable Personal Consumption. As a small business advisor in a rural community, he could only affect the economic systems where he could wield his purchasing power. Yet those economic systems are adjacent to others, which are adjacent to still others, until the systems culminate in the UN SDGs that he valued, namely *UN SDG 10: Reduce Inequalities Within and Among Countries* and *UN SDG 12: Responsible Consumption and Production*. This created a red thread between personal consumption choices and global sustainability goals.

Second, the author made a list of the economic subsystems available for him to participate in, such as local, second-hand, eco-friendly stores, discount brands, and so on. Personal sustainability wasn't just doing without; it was actively choosing in which economic subsystems to participate. Third, the author prioritized his options by using the *avoid, minimize, and manage* framework of managing climate risk. This helped him redirect his spending power in support of his sustainability goals.

The rubric is presented in Figure 5.1. It is a prioritized list of consumption strategies within nested subsystems of progressively greater sustainability impact and wealth accumulation. The rubric is not a formula, where every person will arrive at the same conclusion about a given purchase. Instead, it is a decision guide with a gradient of consumption options that consumers can reference to make the most beneficial choice for their given circumstance and available wealth. The lower the number, the smaller the feedback loop, the fewer resources are impacted in an unsustainable way, and the less spending power is contributed to global wealth accumulation.

AVOID impact by not requiring production:

1. **Don't need it**: Buy nothing. Do without. Do I need it?

2. **Find it free**: Borrow, barter, trade. Buy Nothing Groups.

3. **Make it**: Repurpose, build from scratch, grow it.

MINIMIZING impact by extending the life of existing products or redirecting financial resources to local economies:

4. **Buy used**: Thrift stores, garage sales, Craigslist, etc.

5. **Buy local**: Local farmers and food systems, local craftspeople, locally owned businesses, etc.

6. **Buy from independent or underserved populations**: Independent, women owned, minority owned, LGBTQIA+ owned, etc.

MANAGING impact and reducing economic contribution to large corporations when purchasing from mainstream sources:

7. **Buy sustainably sourced goods**: Places less stress on the environmental systems by buying from sustainable sources or buying goods that have a long useful life.

8. **Buy from discounted**: Intercept waste stream Sierra Trading Post, Big Lots, etc.

9. **Buy durable from commercial retailer**: Amazon.com, big box stores, etc.

Figure 5.1   The author's rubric for sustainable personal consumption

Examples of the Rubric for Sustainable Personal Consumption

When teaching the rubric in class, the author uses the following examples. The authors encourage readers to use their own examples to illustrate the rubric in action.

One of the authors uses this rubric for his weekly food shopping habits. First, he is fortunate to live in a rural area, where he plants a vegetable garden and where his neighbors will share their harvests. These constitute his *avoidance* strategies. Second, with ready access to farms with farm stands, he buys as much as possible, eating what's produced locally. This is where he typically buys his eggs, milk, greens, and vegetables that are in season. For items that can't be purchased directly from local farmers, he shops at a locally owned grocery that also operates two farms. This is where the author buys imported vegetables that can't be found at the farm stand, as well as bulk and packaged goods. These constitute his *minimization* strategies. Third, the author, who is on a budget, *manages* his impact when he cannot afford to purchase at the local grocery. He will hit the locally owned off-price surplus discount franchise, which carries a mix of fresh and packaged foods. When buying bulk and packaged foods, he tries to buy small, local, or organic so as to still make the purchase as sustainable as possible. Finally, for those remaining items that he needs and hasn't found, he shops at a corporate chain supermarket. In this manner, he is able to contribute a significant portion of his purchasing power to local, small businesses before supplementing with purchases from corporations that depend on the global supply chain.

As another example, one of the authors is a committed bargain shopper for clothes. Certain necessities, such as daily clothes, can be purchased at thrift stores or the local brick-and-mortar location of a national off-price retailer. When he needs to buy items that can't be found locally or sustainably produced, such as work suits, he has saved enough money to purchase durable goods and fashion that can last a long time.

There is no one right or wrong purchase choice for all consumers, or even for any one consumer. The same person may have different optimal consumption strategies at different times based on their available resources. The rubric is flexible to guide consumers to their optimally sustainable consumption choice in any given context. For example, one of the authors lives in a dense urban environment and has not owned a car since moving there (avoidance strategies). He prioritizes his transportation needs by walks to his local shops, taking public transportation for further distances and only occasionally renting a car for needed trips (minimization strategies). The author teaches at several schools across the US and his classes are mostly online, reducing the need to travel. However, on occasions, classes are in person and he needs to travel, creating a large carbon footprint. The mobility solution to flying is limited to the major airline carriers, an unsustainable activity. Therefore, whenever the author travels, he consciously buys his goods and services from *local producers*, so that the products he consumes while traveling don't need to generate a large carbon footprint or further enrich large corporations.

Observations of Applications of the Rubric for Sustainable Personal Consumption

The most sustainable choice a person can make with financial power is to not wield it in the first place. The Intergovernmental Panel on Climate Change (IPCC) has observed that global consumption of goods and services is a key driver to climate change. Dematerialization of the economy is therefore crucial to meeting sustainable development goals (IPCC, 2014). In essence, one cannot "buy" their way to a sustainable future when the act of buying is itself unsustainable. *Avoidance* strategies are those where one can either choose to do without or acquire assets through non-monetary exchanges, such as making it or borrowing it. This avoids the supply chain directly. Examples of the authors' practice include living near work to avoid owning a car or maintaining a vegetable garden.

The second most sustainable choice one can make is to *minimize* consumption. In the popular press, there has been a recent increased interest in decluttering one's home (Kondo & Hirano, 2015). Several point to how decluttering one's life can also lead to better sustainability outcomes, reducing what one purchases (Borunda, 2019; Chung, 2019; Lund, 2019). When purchases need to be made, they can also be done to minimize contributions to the mainstream economic engines. The individual can direct capital toward the local economy, such as microenterprises, cooperatives, and women- and minority-owned businesses. The individual could also participate in the circular economy by buying second-hand products. The authors perform this by buying from local producers, second-hand shops, and being intentional to only make purchases when necessary.

Finally, for products that cannot be sourced using any of the previous methods, one can *manage* their purchasing power by supporting best practices within the dominant firms. Companies that have strong sustainability practices are preferrable to those that don't. These include sourcing more sustainable or beneficial options or buying at off-price retailers as a way of diverting imperfect or surplus products from landfill. In these circumstances, it may be best to prioritize the durability of the product, maximizing the utilization of the product's embedded energy over its lifetime. Examples of the authors' practices include buying durable cook wares for long-term use and buying from sustainable brands.

SUMMARY

In summary, the methodology of Tracing the Red Thread demonstrates the interconnections of complex systems in an approachable manner. One complex system is depicted by the Energy Literacy Matrix, a socio-environmental-economic framework depicting the balances among the energy supply chain and societal needs. The nine cells represent subsystems, each with their own goals and constraints. The Energy Literacy Matrix helped one of the authors create a balanced MBA curriculum on sustainable energy that showed students the relationships among seemingly disparate system components. Two key lessons from the matrix are to prioritize decisions based on context and minimize climate-related risks using an avoid, minimize, and manage framework.

These lessons inspired a former student to create a Rubric for Sustainable Personal Consumption, applying the key learnings from class to everyday life. The rubric was developed based on a realization that there are a plethora of retail options available to minimize resource impact while maximizing support of the local economy. The rubric helps individuals make the best purchasing decision for their context by prioritizing the systems in which they can participate. The rubric consists of a series of nested feedback loops, each of which depends on a larger system to deliver the product or services needed by the consumer. By depending on only the systems the individual requires, the person can maximize their outcomes while minimizing their impact.

In essence, if all systems are interconnected, one starts making an impact by influencing the systems they touch.

REFERENCES

Borunda, A. (2019). *Marie Kondo helps declutter homes: What does that mean for plastic waste?* National Geographic, accessed on August 1, 2020 at https://www .nationalgeographic.com/environment/2019/03/marie-kondo-plastic-trash/

BPA. (2020). *Bonneville Power Administration*, accessed on August 3, 2020 at https:// www.bpa.gov/Pages/home.aspx

Burrows, E. G., & Wallace, M. (2000). *Gotham: A history of New York City to 1898.* New York, NY; Oxford: Oxford University Press.

Casillas, C. E., & Kammen, D. M. (2010). Environment and development: The energy–poverty–climate nexus. *Science, 330*(6008), 1181. https://doi.org/10.1126/science .1197412

Chung, S.-W. (2019). *Is Marie Kondo the answer to our environmental woes?* Greenpeace International, accessed on August 17, 2020 at https://www.greenpeace .org/international/story/20317/is-marie-kondo-the-answer-to-our-environmental -woes/

Erickson, L. E., & Jennings, M. (2017). Energy, transportation, air quality, climate change, health nexus: Sustainable energy is good for our health. *AIMS Public Health*, *4*(1), 47–61. https://doi.org/10.3934/publichealth.2017.1.47

Finke, T., Gilchrist, A., & Mouzas, S. (2016). Why companies fail to respond to climate change: Collective inaction as an outcome of barriers to interaction. *Industrial Marketing Management*, *58*. https://doi.org/10.1016/j.indmarman.2016.05.018

Grint, K. (2010). Wicked problems and clumsy solutions: The role of leadership. *The New Public Leadership Challenge*, 169–186. https://doi.org/10.1057/9780230277953

Haley, B., Kwok, G., & Jones, R. (2016). *Deep decarbonization pathways analysis for Washington State* (April), accessed on July 15, 2020 at http://www.governor.wa.gov/sites/default/files/Deep_Decarbonization_Pathways_Analysis_for_Washington_State.pdf

IPCC. (2014). Sustainable development and equity. In: *Climate Change 2014: Mitigation of climate change: Contribution of Working Group III to the Fifth Assessment Report of the Intergovernmental Panel on Climate Change*. https://doi.org/10.1017/cbo9781107415416.010

Jevons, W. S. (1866). *The coal question: An inquiry concerning the progress of the nation and the probable exhaustion of our coal mines* (2nd Edition). London: Macmillan.

Jia, J. Y. (2018). *Master Lecture Slides: Repository of lecture notes developed for sustainable energy, strategy and climate change curriculums*, accessed on August 20, 2020 at https://github.com/JJia/Master-Lecture-Slides

Jia, J. Y. (2020a). *Levers for change podcast*, accessed on August 20, 2020 at www.leversforchangepodcast.com

Jia, J. Y. (2020b). *The corporate energy strategist's handbook*. New York, NY: Springer International Publishing. https://doi.org/10.1007/978-3-030-36838-8

Jia, J. Y., & Crabtree, J. (2015). *Driven by demand: How energy gets its power.* Cambridge: Cambridge University Press. https://doi.org/10.1017/CBO9781316221778

Kondo, M., & Hirano, C. (2015). *The life-changing magic of tidying: The Japanese art*. London: Vermilion.

Lucas, K., Renn, O., Jaeger, C., & Yang, S. (2018). Systemic risks: A homomorphic approach on the basis of complexity science. *International Journal of Disaster Risk Science*, *9*(3). https://doi.org/10.1007/s13753-018-0185-6

Lund, K. (2019). *Environmental lessons from Marie Kondo's "Tidying Up"*. Student Environmental Resource Center, U.C. Berkeley, accessed on August 14, 2020 at https://serc.berkeley.edu/environmental-lessons-from-marie-kondos-tidying-up/

MunichRE. (2017). *Natural disaster risks: Losses are trending upwards*, accessed on August 10, 2020 at https://www.munichre.com/en/risks/natural-disasters-losses-are-trending-upwards.html#-1624621007

Öberg, C., Huge-Brodin, M., & Björklund, M. (2012). Applying a network level in environmental impact assessments. *Journal of Business Research*, *65*(2). https://doi.org/10.1016/j.jbusres.2011.05.026

Rittel, H. W. J., & Webber, M. M. (1973). Dilemmas in a general theory of planning. *Policy Sciences*, *4*(2). https://doi.org/10.1007/BF01405730

Savageau, A. E. (2013). Let's get personal: Making sustainability tangible to students. *International Journal of Sustainability in Higher Education*, *14*(1). https://doi.org/10.1108/14676371311288921

Thoman, R. L., Richter-Menge, J., & Druckenmiller, M. (2020). *NOAA Arctic report card full report | enhanced reader*. https://doi.org/10.25923/MN5P-T549X

UNEP. (2009). *The global state of sustainable insurance: Understanding and integrating environmental, social and governance factors in insurance*, accessed on August 14, 2020 at https://www.unepfi.org/fileadmin/documents/global-state-of-sustainable-insurance_01.pdf

Verweij, M., & Thompson, M. (2006). *Clumsy solutions for a complex world. Governance, politics and plural perceptions*. Basingstoke: Palgrave Macmillan.

Watson, A. Y., Bates, R. R., & Kennedy, D. (Sponsored by the Health Effects Institute) (1988). *Air pollution, the automobile, and public health*. National Academies Press. http://ebookcentral.proquest.com/lib/oxford/detail.action?docID=3377494

Waughray, D. (2011). *Water security: The water–food–energy–climate nexus*. Washington DC: Island Press.

6. Creating connections for progress toward sustainability

Kevin D. Carlson and John H. Grant

This chapter provides a synthesis of ways in which actions from two personal sustainability journeys converged during a business college's transition toward better collective understandings of sustainability, as part of the broad context for improving the human condition (IHC). The processes of search and discovery to identify the roles the individuals might assume in helping with connections throughout the Pamplin College of Business are outlined, as the college seeks to play a meaningful role in addressing some of the most significant challenges to students' futures. The journey includes collaboration across the University, as members of the physical and social sciences also searched for connectivity among concepts, research data and project implementation.

The roles of business schools and their leaders, both officials and other supporters, have been debated for decades. Two stimuli for such discussions came from a coalition of interested faculty from several countries who developed a collaborative initiative that led to the publication of *Management Education for the World* (Muff et al., 2013) and editors interested in future research agendas (George et al., 2015). More recently, the Association to Advance Collegiate Schools of Business (AACSB, 2021) has been updating its accreditation standards with the objective of connecting them more closely to the current contexts of business organizations in the global economy (focus on Standard #9). In particular, many faculty members around the globe have been criticized for failing to "practice what they preach" in terms of sustainability, particularly regarding attendance at conferences and meetings (Higham & Font, 2020). From another perspective, a set of authors analyzed more closely faculty members' personal actions toward *sustainability* (Kanashiro et al., 2020) and summarized them, to illustrate ways in which faculty can set positive sustainability examples for students, colleagues and their communities more broadly. Such steps in the sphere of sustainability provide the context for the activities, both in terms of personal actions and institutional systems changes, as discussed herein.

DISCOVERIES FROM ANALYSES AND SYNTHESIS

Virginia Tech (or VT) has a long history of research and service toward IHC. (Older readers may be more familiar with the name Virginia Polytechnic Institute and State University, in use before VT's rebranding several years ago.) VT's sustainability journey is punctuated by occasional stories that captured broader national recognition. Such notable events include an emeritus biology professor's 2010 article highlighting increasing "threats to the biosphere from interacting global crises" (Cairns); more recently the work of university faculty to highlight system-wide urban "water crises" (Adams & Tuel, 2016) as well as COVID-related research on the dynamics of aerosols (Irby & Nelsen, 2020) have attracted national attention.

Many additional efforts toward IHC occur and are recognized primarily within the university. Such activities include the president's issuing and recently updating the university's "climate action commitment," annual Earth Week celebrations, a vibrant College of Natural Resources and Environment, and several environmentally focused research centers, e.g., its Center for Coastal Studies. As the authors spoke with selected faculty in several departments across two of VT's major campus locations, they discovered both externally and internally funded sustainability projects ranging from water quality enhancements to advanced electrical system efficiency programs. The second author recognized that far more was being done at Virginia Tech, as he had recognized at other such institutions, that "falls below the radar" of the broader community—only recognized by those directly involved. Lack of awareness of disparate efforts has undoubtedly limited possible synergies across the university. Such missed opportunities for collaboration can dampen momentum for new initiatives toward IHC.

Across the Pamplin College of Business, discussions about sustainability, climate change, and IHC had been a continuing undercurrent in faculty research and education, but had never gained the momentum to achieve "consciousness" in the college's planning. That changed during the most recent strategic planning cycle. The first author was involved with an extensive environmental analysis that led to more systematic consideration of a wider range of environmental influences over a longer than typical planning horizon (Hawken, 2017; Romm, 2018; Crist, 2019; Wunder, 2019). While the implications of analytics and artificial intelligence, cybersecurity, the future of health care and online learning were recognized, the dramatic potential for climate change effects were underestimated. When the implications of the "100-year life" (that steadily increasing lifespans suggest that today's 20-year-olds have a 50% chance of living to 100) framed the present and growing challenges facing students during their uncertain futures (Gratton & Scott, 2017), further actions seemed necessary. As ecological shifts were matched with VT's motto *Ut Prosim* ("that I may serve"), a new reality provided a dramatic call to action for which appropriate responses were both unclear and systemic (Strom, 2015).

The breadth of what it could mean to "improve the human condition" became apparent during the development of Pamplin's new strategic plan that, for the first time, included the phrase as a college "pillar," i.e., an area of research and practice that reflects critical societal needs or important areas of emerging practice that should be targets for investment by the college. Through discussions among faculty, students, advisory board members and administrators, the college began to recognize the myriad emerging threats to the quality and sustainability of human life and livelihood facing individuals, families, economies and the planet during the lifetimes of students. The college also recognized that identifying and understanding these systemic problems would be critical to developing and deploying meaningful solutions, and well-prepared business graduates could be important partners in these efforts. The college also recognized there was much it did not know if they were to make such changes successfully.

Faculty members' personal assessments of external global sustainability reports, e.g., those from the IPCC (Intergovernmental Panel on Climate Change), the Paris Accord of 2015, the *Laudato Si* encyclical on ecology of 2017 from Pope Francis, and "The Planetary Emergency Plan" of the Club of Rome, left them with varying senses of urgency. As educators, several of the faculty had begun to take individual actions to broaden the analytical frameworks in some of their courses to include non-human concepts such as carbon footprints and sea-level rise, while at the same time recognizing the growing importance of racial, gender and racial justice issues. A few faculty members had been following the PRME (Principles of Responsible Management Education) organization as well as the Organizations and the Natural Environment (ONE) Division of the Academy of Management, and others were aware of VT's Center for Coastal Studies group that drew researchers from across multiple campus locations. At the same time, other faculty felt that their traditional models implicitly allowed for these variables if students chose to include them for their particular projects, firms or industries of interest, so significant changes or adaptations probably would not be needed.

During the spring of 2020, Pamplin retained an external consultant and former university professor with extensive "sustainability" experience to help several in Pamplin better understand what IHC might mean in terms of the tangible, natural ecosystems supporting human civilization. A number of facilitated discussions helped highlight current perspectives on what IHC could mean in Pamplin and across Virginia Tech. Three insights emerged for the authors from these discussions. First, Pamplin already had a significant number of faculty who saw their research, teaching or service work as directly related to improving the human condition, and a larger group of faculty members could define their work as contributing in some way to these efforts.

Second, many faculty and students were passionate about improving the human condition, but their individual efforts reflected idiosyncratic approaches that, on the surface, seemed difficult to reconcile. At the time, Pamplin had no common understanding of the conceptual boundaries of "the human condition" or the most meaningful approaches to improving it.

Third, engagement with IHC looked fundamentally different in Pamplin than in most of VT's other colleges. Business faculty and students often viewed improving the human condition through the lens of commercial organizations. On the other hand, researchers elsewhere on campus conducted basic research regarding the nature of health, greenhouse gasses, water quality, infectious diseases, poverty and many other factors that influence the human condition. These differences in focus and perspective created challenges, but also opportunities across the university system.

While business students developed strong problem-solving skills for addressing ambiguous and uncertain problems, they often defined those problems rather narrowly, increasing the potential that their solutions created externalities because negative consequences remained unforeseen, lurking just outside problem frames. Depletion of natural resources, destructions of habitat, pollution and child labor were such examples. On the other hand, faculty in the natural and physical sciences struggled to communicate threats to the human condition in ways that would persuade business leaders to act. Pamplin had much to learn from colleagues in the "hard sciences," and Pamplin had much to offer that could increase the positive effects of the basic sciences on the human condition.

At the same time that many factors were encouraging individuals to move forward with personal sustainability actions, there were countervailing forces from political ecosystems both inside and outside of many universities. Hence, administrative officials often had incentives to downplay "greening" initiatives or to proceed in "stealth" fashion, as described by Starik (2015). While many potential donors have given and promised millions of dollars to support sustainability, others have offered universities resources to "silence" professors who are public advocates regarding their sustainability concerns (Lockwood, 2017).

LEVELS OF PERSONAL ACTIONS

Despite the fact that many global trends pertaining to sustainability and the human condition were continuing to move in the wrong direction, the authors and many Pamplin faculty had taken *personal steps* to set positive examples toward sustainability. Such actions range from relatively minor to quite significant, e.g.:

- purchasing small items like shirts and ties with patterns chosen to stimulate discussions of SDGs like #12 (Responsible Consumption) or #15 (Biodiversity)
- choosing to have a yard certified as a vegetation and animal habitat by the National Wildlife Federation for supporting SDG #15 (Biodiversity)
- placing sustainability books in conspicuous locations in one's house, so visitors could feel free to initiate such discussions of feasible actions, e.g., those from Project Drawdown (Drawdown, 2020; Hawken, 2017)
- riding bicycles and public transportation to reduce GHGs as well as add to exercise routines. Other faculty had installed solar panels that were connected to local energy grids. All of these actions further supported SDG #7.

On a larger scale, faculty members can take direct actions to reduce their carbon debts by investing in the work of others or changing their lifestyles, e.g.,

- contributing by the authors of thousands of dollars over three decades to numerous environmental organizations with global operations, including WRI (World Resources Institute), NRDC (Natural Resources Defense Council) and the Environmental Defense Fund (EDF)
- donating substantially to certified carbon offset providers like Native Energy, the Colorado Carbon Fund and the Carbon Trust; such investments can frequently be woven into discussions with friends and extended family members of both short-term as well as long-term benefits to the human condition
- several faculty members, as well as the second author, had purchased hybrid or electric automobiles to reduce emissions and to provide segues to "travel discussions" involving the various ecological trade-offs in such decisions, whether for business or pleasure
- on an even larger scale, a "passive solar" house had been purchased by the second author to help mitigate GHGs, even before SDGs like #7 (Energy) had been specified.

Table 6.1 provides a summary of the levels of actions and links them to various SDGs. Our presentation here begins with *personal* actions at the bottom, so one can imagine building upward and broadening the scope of involvement to the global level.

From an *organizational* perspective, the authors and several other faculty had made presentations on a range of "sustainability topics" to academic audiences where hundreds of people had heard about possible solutions to

Table 6.1   *Personal actions by levels, types and SDGs*

Levels	Actions by Type and SDG links		
	Human*	Non-human**	SDGs
Global or socio-political "movement"	Presentations at international conferences	CO_2 reductions via donations for carbon offsets	# 13 – Air quality # 4 – Education
Regional	Donations to food banks	Hybrid vehicle purchase	#2 – Nutrition #12 – Production and consumption responsibly
Industry or organization	Letter to editor re. Carbon offset costs omitted from stadium budget. Biodigester design	University curricula Climate Change books to faculty LLC collaborative design Biodigester system evaluation	# 4 – Education # 4 – Education # 14 – Food waste # 12 – Production and consumption responsibly # 7 – Energy
Individual persons	Conferences attended for advocacy networking and learning Cycling to work Clothes with theme patterns reflecting SDGs	Syllabus preparation for students' Asia sustainability site visits Wildlife habitat in backyard Passive solar house purchased	# 6 – Water # 4 – Education # 15 – Biodiversity # 7 – Energy # 13 – Climate # 4 – Education

Notes: * Human refers to authors' personal participation. ** Non-human includes actions via money and technology.

"natural capital" measurement systems, "green GDP" links to ecological realities, the roll-out of Future Earth and related topics. Pamplin provided copies of *Climate Change*, 2nd ed. (Romm, 2018) to all college faculty and staff to further increase awareness of climate change concerns and to promote broader conversations regarding the role of sustainability in business education for the 21st century, all in support of SDG #4 (Education).

The second author wrote letters to editors, including a very unpopular one proposing the inclusion of purchased "carbon offsets" in the budget for a new football stadium at Colorado State University. The suggestion was not very popular with the athletic department or its allies, many of whom were being pressed by the university to increase their pledges for the project! On the other hand, overseas executive MBA trips to Europe and Asia, which were conducted with foci on sustainability in both the private and public sectors, were received more favorably by the students (SDG #4). Such trips were recognized by many as a double-edged sword, with knowledge of "sustainability" being enhanced, while at the same time additional GHGs from extensive travel were polluting the atmosphere (SDG #7). The authors realize that such trade-offs are quite common in the personal sustainability field, so decisions need to be made with both the longer time horizon and larger social system in mind.

The second author became involved with another sustainability system with a circular economy facet in the form of a biodigester (anaerobic) project aimed at keeping food waste out of landfills, while using methane (CH_4) to generate electricity. This project exemplified the reciprocal interactions among the human and non-human elements of a project consisting of multiple components. Many alternative project scopes and sites were explored, and each required the team to consider different technical systems for collecting deteriorating food and various levels of training for potential staff members. If the facility could be located near a major university, then the biodigester could be used to help educate the college community as well as to provide a research site for those studying renewable energy systems! (SDGs # 2 and 7.)

To facilitate potential transitions to a broader level of analysis, the first author encouraged faculty consideration of possible collaborations, as both authors had witnessed at other locations. By encouraging faculty to become better prepared for discussions with potential donors, awareness of the breadth and depth of skills within Pamplin and across Virginia Tech could help funding proposals become increasingly responsive and probably more creative. Any such collaborative actions were viewed as adding to the organizational learning within the University.

From a "movement" or *global* vantage point, the second author had previously supported "clean cookstove" development efforts to help save vegetation, reduce GHGs, and permit more educational participation in villages across several developing countries using a systems perspective. While the engineering aspects of such stoves were challenging enough, the procurement of financing, supply chain development and cultural adaptation had been even greater barriers to sustainability progress (Clean Cooking Alliance, 2015). Entrepreneurial concepts led to the connection of micro-financing to the thermal-electric generation (TEG) and storage capabilities that could make such stoves self-financing, given contiguous electrical systems and markets.

One of the authors had made several presentations at conferences of an international organization that met in various countries in order to accommodate "visa restrictions" on faculty from certain countries deemed hostile to the USA. One of these presentations led to a paper designed to improve organizational performance measurement to address sustainability issues (Grant, 2008). Rather than purchasing carbon offsets for travel, the presenter paid several hundred US$ to cover the registration fee for a faculty member from a coastal LDC, so that person could help leverage further climate education in his or her home region.

The second author of this chapter had spent several months on a part-time, *pro bono* basis assisting with the ramping-up of the Future Earth organization based upon several sustainability research associations around the globe (Future Earth, 2020). Discussions with the Executive Director sometimes turned to the challenge of balancing the physical (non-human) and political (human) aspects of "sustainability," as "partnerships" were being encouraged (SDG #17). While the future of civilization(s), including its/their size, may be heavily dependent on the basic physical foundations of the biosphere, the political realities of many impoverished countries are that human institutions need a lot of rather immediate support to avoid systemic collapse (consider the Society level).

While levels of analysis provide one way of viewing sustainability activities, the reader can also think about connecting from the foundations of the natural world upwards through linkages among human institutions, as with SDG #17 or partnerships.

EVOLVING FRAMEWORK FOR CONNECTING ACADEMIC COMMUNITY MEMBERS

The disparate views of ways to improve the human condition made it challenging for the Pamplin leadership to choose among areas for academic investment. As faculty and students explored new avenues for improving the human condition, a common framework for understanding and integrating these efforts was thought to be useful. One such integrating framework had emerged from a unit of Stockholm University. A team led by Johan Rockstrom, then of the Stockholm Resilience Centre (SRC), focused on categorizing and developing linkages among the UN's 17 SDGs. In an effort to integrate these goals, the SRC created a static three-tiered "wedding-cake" model (Rockstrom & Sukhdev, 2016).

For our purposes, we have modified the "cake" into a more dynamic "spinning-top" model, as illustrated in Figure 6.1. This visualization was thought to be more effective in communicating the dynamic connections among the SDGs within a research-oriented university. Think of research, teaching and outreach as moving independently similar to many university departments, but with opportunities for "alignment" when administrators or entrepreneurial faculty promote connections and linkages.

In Pamplin's model, the *bottom* level (dark gray) represents the "biosphere," a term borrowed from Rockstrom and Sukhdev, rather than a more encompassing "natural environment" phrase. This is where concerns about the quality and sustainability of soil, water, atmosphere and biodiversity reside. SDG numbers 6, 13, 14 and 15 are logically assigned to this level. These SDGs provide the "building blocks" upon which other goals rest and gain sustenance.

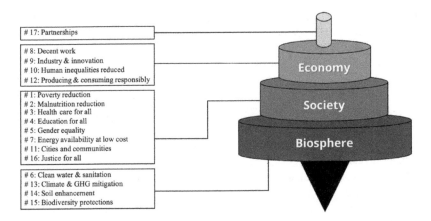

*Figure 6.1   Spinning-top model for improving the human condition
                (IHC), including SDGs*

The *middle* level (medium gray) addresses "societal" concerns, e.g., food insufficiency and poverty, income and wealth disparity, social and political instability, inadequate health care, injustice and insecurity, gender inequality and educational inadequacy. SDG numbers 1, 2, 3, 4, 5, 7, 11 and 16 comprise this intermediate level. A society's progress on these various SDGs determines the immediate contexts within which individual managers, employees and investors gain guidance about the treatment of other humans in terms of physical existence and notions of justice and fairness.

The *upper* level (light gray) of the model addresses issues related to "economies." We can view an economy's infrastructure as the traditional physical roads, bridges, internet wires and towers, and harbors, that allow goods and services to move from producers to consumers to sellers. In addition, the Economy includes the intangible "organizational systems" of incentives, regulations, penalties and "equal opportunity." SDGs numbers 8, 9, 10 and 12 interact at this level. It is in this region that much business school activity typically occurs.

SDG number 17 (Partnerships) is the "spinner" at the apex of the model that can be viewed as providing the energy to initiate and sustain the connections among many of the other SDGs. It is perhaps in the development and maintenance of fruitful partnerships that the greatest challenges to social skills in the pursuit of other SDGs are found (Jones, 2020).

An important feature of this framework is its hierarchical depiction of human condition concerns. Economic concerns are important, but they depend on an underlying set of societal factors to be present in order to sustain and then seek to optimize economic outcomes. Likewise, unmitigated deficiencies in biospheric factors create challenges for assuring both effective societies and positive economic outcomes. The spinning-top model reminds us of the need for balancing at each level of the framework as well as demonstrating opportunities for independent alignment among activities connected to various sets of SDGs.

In order to help students benefit early in their college experiences from the rich array of VT's learning opportunities, the campus had previously initiated student housing opportunities for entrepreneurship, business analytics and artistic/cultural skills. The successes of these units led the first author and the IHC implementation staff to ask, "Should we plan an LLC for Environmental Sustainability (ES) or some similar theme?"

NEXT CONNECTIONS: LIVING–LEARNING COMMUNITY?

One of the most significant organizational-level actions that emerged from discussions about IHC was the expansion of Living–Learning Communities (LLCs) for undergraduates across the university as a means of helping them learn "beyond the silos (or 'boundaries')" of their individual colleges and departments. LLCs were seen as vehicles for helping students achieve the aspirational aims of IHC, including accelerating student achievements in scientific, societal and business foundations for activities that would support SDGs in a variety of ways. As a result of informal discussions among a few faculty and Associate Deans, an idea emerged for a new LLC that would include Pamplin students along with those from the College Architecture & Urban Studies (CAUS). About 25% of the students would be drawn from any other college as long as they expressed some interest in the objectives of the new LLC and were willing to participate in such "learning networks" (Hull et al., 2020).

Although a few different physical designs and locations had been considered by the university through late 2020, a specific set of decisions had not been made by early 2021. This was due in part to the uncertainties on several fronts generated by the COVID-19 pandemic, which had been impacting universities for nearly a year and were expected to continue to do so for at least many more months.

In summary, this chapter has integrated the authors' personal experiences applying sustainability concepts and extending them into the context of a major research university. The two faculty members and many colleagues drew from a wealth of experiences studying, acting upon and communicating sustainability concepts at many levels, from personal to global. We believe the spinning-top framework can be useful for integrating a university's disparate efforts and identifying areas where further thinking and contributions to sustainability progress can be made. Further, we believe this framework can help Pamplin identify relevant partners, attract further faculty contributions, broaden student talent and attract new sources of funding. We hope it will help inspire similar efforts across other universities, because accelerated attention to the SDGs and other sustainability matters is urgently needed.

REFERENCES

Adams, M. & Tuel, J. (2016). Fighting for Flint. *Virginia Tech Magazine*, Spring.

Association to Advance Collegiate Schools of Business (2021). Standards overview, at www.AACSB.edu/accreditation/standards (accessed January 8, 2021).

Cairns, J. Jr. (2010). Threats to the biosphere: 8 interactive global crises, *Journal of Cosmology*, 8 (June), 1906–1915.

Clean Cooking Alliance (2015). Five years, at CleanCookingAlliance.org/resources/reports/FiveYears.html (accessed January 10, 2021).

Crist, E. (2019). *Abundant earth: Toward an ecological civilization*. Chicago, IL: University of Chicago Press.

Drawdown (2020). Drawdown, at www.Drawdown.org (accessed September 8, 2020).

Future Earth (2020). Our history, at FutureEarth.org/about/history/ (accessed December 28, 2020).

George, G., et al. (2015). The management of natural resources: An overview and research agenda. *Academy of Management Journal*, 58(6), 1595–1613.

Grant, J. (2008). Organizational performance in an interdependent world, in M. A. Rahim (Ed.), *Current topics in management* (13, 35–58). New Brunswick, NJ: Transaction Publishers.

Gratton, L. & Scott, A. (2017). *The 100-year life: Living and working in the age of longevity*. London: Bloomsbury.

Hawken, P. (Ed.) (2017). *Drawdown: The most comprehensive plan ever proposed to reverse global warming*. London: Penguin.

Higham, J. & Font, X. (2020). Decarbonizing academia: Confronting our climate hypocrisy. *Journal of Sustainable Tourism*, 28(1), 1–9. https://doi.org/10.1080/09669582.2019.1695132

Hull, R. B., Robertson, D. P. & Mortimer, M. (2020). *Leadership for sustainability: Strategies for tackling wicked problems*. Washington, DC: Island Press.

Irby, S. & Nelsen, E. (2020). Spurred by COVID-19, researcher Lindsey Marr evaluates efficacy of sterilized N95 respirators, alternative materials. *Virginia Tech News*, May 19.

Jones, R. G. (2020). *The applied psychology of sustainability*, 2nd ed. New York, NY: Routledge.

Kanashiro, P., Starik, M. & Rands, G. P. (2020). Walking the sustainability talk: If not us, who? If not now, when? *Journal of Management Education*, 1–30. DOI: 10.1177/1052562920937423

Lockwood, J. (2017). *Behind the carbon curtain: The energy industry, political censorship and free speech*. Albuquerque, NM: University of New Mexico Press.

Muff, K., Dyllick, T., Drewell, M., North, J., Shrivastava, P., & Haertle, J. (Eds.) (2013). *Management education for the world: A vision for business schools serving people and planet*. Cheltenham, UK and Northampton, MA, USA: Edward Elgar Publishing.

Rockstrom, J. & Sukhdev, P. (2016). How food connects all the SDGs, at www
.StockholmResilience.org/research/research-news/2016-06-14-How-food-connects
-all-the-SDGs.html (accessed January 12, 2021).

Romm, J. (2018). *Climate change: What everyone needs to know*, 2nd ed. Oxford;
Oxford University Press.

Starik, M. (2015). The downside of (and antidote to) unhealthful stealth sustainability. *Organization & Environment*, 28(4), 351–354. https://doi.org/10.1177/
1086026615623589

Strom, D. (2015). *Systems thinking for social change: A practical guide for solving
complex problems, avoiding unintended consequences, and achieving lasting results.*
White River Junction, VT: Chelsea Green Publishing.

Wunder, T. (Ed.) (2019). *Rethinking strategic management: Sustainable strategizing
for positive impact.* Cham, Switzerland: Springer.

7.   Cultivating the ecological imagination
Billy Friebele[1]

My research significantly impacts my teaching, service, and sustainable actions at home. Likewise, my artwork is influenced by my students' care for the environment. Cultivating a sustainable lifestyle on all three of these fronts requires time and energy. It also necessitates a shift away from acting solely through the lens of speed, production, and profit, towards a more wholistic, and patient mindset.

RENEWABLE ENERGY

I design creative projects that integrate sustainable practices, adopting an approach that focuses on process over product and slow growth over rapid returns. While it can be argued that art does not make a direct impact, my goal is to raise awareness about the potential of sustainable energy sources. There are a number of artists who use solar cells in their artwork, primarily for light art installations that are visible at night. I do not claim that my sustainable art projects are entirely unique, but kinetic solar projects are less common, and I utilize motion as an element of surprise to catch the viewer off guard with the goal of calling them action.

In order to gain technical knowledge, I frequently propose projects that require new skills. For example, I applied to an open call for outdoor sculpture with a proposal for a kinetic artwork powered by small solar cells. The renewable energy produced charged batteries, enabling the artwork to respond to viewers interactively. This proposal was accepted, which forced me to learn a tremendous amount of electrical engineering.

The exhibition theme was introspection and I used cherished objects that mark time and frame our growth as humans, such as trophies from my childhood and my son's play harp, in a kinetic sculpture entitled *Reflection*. This piece integrates a playful quality with renewable energy practices, drawing the viewer in with a simple interaction that leads them to examine a network of connections tracing back to solar cells as a power source.

Two small 2W/6V solar cells charge Lithium-Ion batteries. An infrared sensor detects when a human is present; a microcontroller then triggers a servo motor to rotate an arm to a random position and back to the origin point.

A key suspended from this arm strikes the child's harp and trophy figurines to produce sounds. I coded an Arduino Mini Pro microcontroller to control the servo motor and made several alterations to reduce the power consumption such as cutting a lead on an LED. This sculpture was installed at the Sandy Springs Museum in July of 2018 and functioned outdoors for four consecutive months (Figure 7.1).

Figure 7.1   Photograph #1: Reflection

RENEWABLE ENERGY AT HOME

In an effort to address the SDGs on a slightly larger scale, I purchased a set of six 100W/12V solar panels and a 40A Maximum Power Point Tracking solar charge controller to further experiment with solar energy. I learned about the intricacies of energy consumption by powering electrical devices in my shop with a marine battery charged by solar panels. The charge controller is Bluetooth enabled, allowing me to see how much electricity is flowing from the solar panel and the level of charge in the battery, as well as the amount of energy consumed in real time on an app installed on my phone.

I am currently working to install these panels on the roof of my studio, covering approximately 42 sq. ft. This renewable energy will be used to charge battery-powered tools and equipment. My goal is to phase out gas-powered and electric tools that need to be plugged into the grid. This setup can be expanded by adding batteries and solar panels as needed.

This experiment led me to pay closer attention to the appliances in our house and whether electricity is being consumed efficiently or inefficiently. The simple act of making the invisible visible is powerful, and this is what I try to accomplish with my sustainably focused artwork.

SOLAR SCULPTURE

Building on the knowledge gained from *Reflection*, I created a large-scale (120" × 92" × 40") outdoor kinetic sculpture that uses solar panels to produce sound, entitled *Nero Plays a Fiddle*. I used a windshield wiper motor wired directly to a 100W/12V solar panel to visualize the energy produced. When the sun strikes the solar panel, the motor engages, causing a silicone hand to strike metal upright bass strings, producing a plucking sound. This action refers to the Roman emperor Nero, who allegedly played the fiddle while Rome burned. Through this sculpture I connect this legend with our contemporary climate dystopia.

The tempo of the sound is a direct reflection of the amount of electricity being produced by the solar panel, and this real-time translation is especially evident on a partly cloudy day. When a cloud passes in front of the sun, the motion slows to a crawl, and when the sun reemerges, it quickens. Witnessing this transformation of solar energy makes the process understandable for the public.

I used a second solar panel to charge a 12V/100Ah AGM marine deep cycle battery. The stored 12V electricity flows to a voltage regulator, decreasing it to 6V so it can power an infrared sensor. When a viewer walks in front of the sculpture, the sensor is triggered, closing the circuit and activating a 6V motor, which rubs against an upright bass string. This produces an ominous droning sound like a bowed upright bass. The stored renewable energy and the sensor makes it possible for the viewer to experience the kinetic work on overcast or rainy days when there is little sunlight to power the windshield wiper motor. I experimented with sonar and light sensors, and found infrared sensors draw the least amount of current to operate. Wiring the project without a microcontroller and allowing the sensor to open and close the circuit made it even more efficient.

Nero Plays a Fiddle was installed in the sculpture garden in the Kreeger Museum in Washington, DC from September 2020 through July 2021. The piece is in close proximity to the work of modernist kinetic sculptor George Rickey. My project pays homage to influential artists like Rickey, while pushing outdoor kinetic sculpture to interface with environmental issues, highlighting the potential of sustainable energy. *Nero Plays a Fiddle* addresses the challenges of climate change in the current moment, where environmental threats jeopardize a sustainable future. I use a rusty oil barrel and marine battery as ballast to hold the strings in tension, and this arrangement places the source of our energy dependence in front of the viewer, to force us to look and reckon with our choices (Figure 7.2).

Figure 7.2   Photograph #2: Nero Plays a Fiddle

TEACHING RENEWABLE ENERGY

Whereas my artworks have an indirect relationship with the viewer, through teaching I am able to directly communicate the importance of the SDGs to students and develop projects working with community partners. For instance, at Loyola University Maryland (Baltimore, Maryland), I teach a service-learning course entitled *Public Art*, and our class met with the Community School

Coordinator at Govans Elementary School to discuss a block in the neighborhood that is plagued by lack of lighting, which leads to unsafe conditions for children. Our partners contacted the city about this issue but received no support. We discussed installing a series of solar-powered lamps throughout the block, which would brighten the area while also demonstrating the effectiveness of solar power. Ultimately, we realized that this project is larger in scope than what is possible to accomplish in the timeframe of one semester, but we laid the foundations for executing the project in the future.

Renewable energy projects are time-consuming and complicated, especially when they are installed in outdoor public spaces where liability and safety are a primary concern. I intend to engage students in the creation of sustainable energy projects in the future in a service-learning capacity. It is a process that will take years to unfold, but it is important because knowledge of sustainable energy leads to independence through less reliance on the grid. I am privileged to have access to these resources, and my hope is that sharing this knowledge with those in a less fortunate position will be empowering.

MAKING THE SWITCH TO MORE SUSTAINABLE CONSUMPTION PATTERNS

Beyond sustainable energy, I also investigate waste and consumption in my artistic practice. When my son was born, I realized that the amount of disposable plastic containers used to nurture an infant significantly increased the volume of recyclable materials and trash exiting our house each week. Over a three-month period, I gathered, cleaned, and sorted as many containers as possible. I examined the transparent plastic forms that were used for such a brief period of time. At the time, I was preparing to exhibit a new body of work in Denver, Colorado in a refurbished industrial space with ceilings that are two stories high. I decided to represent the massive amount of material discarded by my family over three months by suspending all of this plastic from the ceiling to the floor in order to create a giant column of plastic. This act highlights the fact that peripheral packaging materials are so ubiquitous that they disappear from our consciousness, yet their impact will last far into the future.

Styrofoam packaging is another harmful material that is ignored and discarded. In the artwork *Disposable Empires* I collected this synthetic material and used it in an interactive sculpture where sensors, motors, and LED lights activate take-out containers and packaging. Animating this harmful material envisions a fantastical dystopian future where Styrofoam gains consciousness and outlives us all (Figure 7.3).

The UN SDG 12, target 12.5, aims to substantially reduce waste generation through prevention, reduction, recycling, and reuse. By embedding interactive components into disposable objects, I render these peripheral ubiquitous materials more visible, which is unique. However, I do recognize that significant impacts in this area will require structural change beyond consumer habits.

Figure 7.3   Photograph #3: Disposable Empires

The gallery where this work was exhibited is named ReCreative Denver and it is "a non-profit organization dedicated to promoting creativity, community and environmental stewardship" (ReCreative Denver, n.d.). In the front portion of the gallery there is a shop that sells recycled art supplies and a number of upcycled items. The definition of upcycle, according to the Merriam–Webster dictionary, is "to recycle (something) in such a way that the resulting product is of a higher value than the original item." This act is essentially the core of my creative process. I collect objects that are disregarded, disposed of, or cast off. I ask myself, what does this material reveal about our culture of consumption? How can I repurpose this item so that it challenges us to think differently?

Repurposing requires a creative or poetic turn. Michel De Certeau examines the way in which we make "innumerable and infinitesimal transformations of and within the dominant cultural economy in order to adapt it to [our] own interests and [our] own rules" (De Certeau, 2011, p. xiv). He refers to this act as *bricolage*. This term was originally coined by French anthropologist and ethnologist Claude Levi-Strauss to describe the act of using "the means at hand" (Levi-Strauss, 1968). There is a lineage of artists, such as the Arte Povera movement, who used common materials to "disrupt the values of the commercialized contemporary gallery system" (Tate Modern Museum, n.d.). A similar term, *rasquachismo* was developed by the Mexican and Chicano art movement to express making "the most with the least" (Gutiérrez, 2017, pp. 184–187). This idea stretches far beyond the art world. Traveling in Latin America and Southeast Asia, I witnessed many creative reuses of common "disposable" materials by necessity. Unfortunately, citizens of wealthy countries like America are often unwilling or unable to see the potential in commonplace materials. Using artistic spectacle, creative reuse is a way to reintroduce these possibilities into our collective imagination provided they are seen by various segments of the public, including students.

RESPONSIBLE CONSUMPTION IN TEACHING AND SERVICE

Bricolage and upcycling are important components in my teaching of studio art – specifically sculpture – in a liberal arts institution. We don't have a foundry, welding tools, or enough space for a dedicated sculpture studio. This limitation leads me to teach three-dimensional art practices in a more sustainable manner. I ask students to bring found objects to class and to repurpose them. We reflect on the cultural values embedded in each object and discuss how we can break them down and reconstruct them in new ways. My intent is to prepare students for a post-graduation creative life where they will be less likely to purchase expensive new materials with a larger carbon footprint. This working process also leads students to understand that ubiquitous disposable materials are a byproduct of capitalist consumer culture and that they can swim upstream rather than accept the model that is given. Anecdotally, one student made a found object sculpture out of disposable materials and later started a campaign against the use of plastic straws on campus.

My classes exhibited found object upcycled sculptures at the Loyola Notre Dame Library and at the Loyola University Maryland symposium for Innovation & Entrepreneurship. These events fostered connections with other faculty and administrators with shared interests. I was invited to teach modules on upcycling at a sustainability teach-in event, for which I developed upcycle projects using plastic water bottles and introduced students to a variety of environmental artists. At each event I worked with a group of five students who were interested in sustainability.

Reflecting on this process, I realize that just taking the time to notice how much trash my family produces and being disturbed by this fact led to a chain reaction of events influencing my creative research, teaching, and service. The simple act of raising awareness among our students – asking them to pause, reflect, and think critically – is vital. In order to "substantially reduce waste generation through prevention, reduction, recycling and reuse," empowering the younger generation is key (SDG 12, target 12.5).

LIFE BELOW WATER

Focusing on plastics consumed by my household led me to notice the proliferation of plastics outdoors. What I used to disregard as peripheral, I could no longer ignore. There is a site that I repeatedly visit with my dog while jogging – a thorny slope that leads to a small rocky shore. As I walk around this site, I find the most unusual debris. Baby shoes, plastic jewelry, a weight set, an oversized container of protein supplements, and bags of clothes. I return to this site through changing seasons and witness the landscape wrapping itself around these deserted artifacts, vines using them for structure, heavy rains washing them into the river and downstream. Plastic bottles bobbing up and down are a frequent, disturbing sight. To fight this, I collect plastic litter as I run to keep it from washing into the river and it amazes me how many bottles repeatedly dot the landscape.

One water sample study conducted in 2019 just downstream, found 441.73 microplastics per liter (MPP/L) in water samples (O'Donnell, n.d.). This number only increases 2 miles further downstream as multiple tributaries join together, where 696.05MPP/L were found.

Watching clear plastic water bottles floating, bumping against rocks, getting stuck in eddies, makes me think of the journey these macroplastics take through local waterways, shedding tiny particles with every encounter. I collected discarded objects on the shore to construct a simple drawing machine, using a plastic bottle as a float to capture the motion of the river (Figure 7.4). This drawing device, *Slosh Cypher #1: Northeast Branch, Anacostia River – Riverdale, MD*, records flowing tides. Dots denote ripples in the water and lines depict smoother passages. Focusing on one site over a prolonged period of time raised my awareness of the proliferation of plastics in this waterway, which flows to larger bodies of water before joining the ocean. The SDG #12 calls us to eliminate plastic usage as much as possible and organize clean-ups, and I continue to take action in this area. This art project impelled me to apply for training to be a citizen scientist and in this role, I will perform stream monitoring and submit findings to a central database (Anacostia Watershed Society, n.d.).

Focusing on the pollution of waterways guided me through an interesting service opportunity on campus. I am the club moderator for the Art Association at Loyola University Maryland, and when students approached to me with the idea to paint a mural on campus, I encouraged them to reach out to the Environmental Action Club at Loyola University Maryland to collaborate. The painting encourages students to refill reusable bottles rather than purchasing new disposable containers.

Figure 7.4   Photograph #4: Examples of Slosh Cypher drawing machine

SUSTAINABLE CITIES AND COMMUNITIES

The work my students undertook on the mural reflects the importance of SDG 14, Life Below the Water. This led me to consider what I can do to positively impact the local watershed. My family decided to replace our concrete driveway with permeable pavers with support from the Prince George's County Raincheck Rebate program, which paid for $4,000 of the cost (Chesapeake Bay Trust, 2021). Many municipalities have comparable programs, and the more impermeable surfaces that are eliminated, the healthier the watersheds will be.

Small projects that make living spaces more sustainable are instrumental in connecting community members who have shared environmental interests. As SDG 17 emphasizes, partnerships are a vital aspect of creating a sustainable future. Public Spaces and Greening Committee (PS&G) is a group of community members from both inside and outside of our university with expertise in landscaping, urban ecology, and environmental design on York Road, a neighboring community to Loyola University Maryland. An impressive project undertaken by PS&G is the Govans Urban Forest (GUF), a plot of land between a pharmacy and a church that the committee cleaned in order to create a small grassroots public nature park (Perl, 2019). They maintain this space by nurturing native plants, creating trails, benches, and signage to identify plants and trees.

Members of the PS&G met with students in my *Public Art* class, explained their mission, and what they hoped to accomplish in the future. Students then worked in collaborative teams to develop detailed proposals, presented these materials to the committee, and listened to feedback. The groups fabricated and installed sustainable public art projects in GUF by the end of the semester.

One student used a laser cutter to craft a new sign for the entrance of the park with the title of the park engraved into cedar boards. Another group created an interactive tic-tac-toe board that children can play with, where the X's and O's are replaced with paintings of native plants, birds, flowers, and insects. A third group designed and installed a series of bee and insect houses to increase the diversity of fauna in GUF.

Students were excited to put their skills to work and make an impact. They reported a positive learning experience due to direct engagement with sustainability issues and interaction with community organizers.

PARTNERSHIP FOR THE GOALS

It is vital for students in this service-learning course to understand the socio-economic divide between the Homeland and Govans neighborhoods, and how this is due to systemic racist policies of redlining, which are discriminatory real estate practices. Students worked in a service-learning capacity at Govans Elementary School, and we met with community leaders who explained the sordid history of segregation of this area. Climate change will affect those with the least resources first, and through these experiences, students learned important lessons about environmental racism.

The intricate web of community support and skill sharing necessary to foster environmental awareness through proactive artistic measures is apparent as we develop the projects. One of the most valuable aspects of the service-learning experience is how, over the course of the semester, students' viewpoints change, from a general feeling that we are in a position to solve the community's problems, to an understanding that lasting change does not occur in the time span of three months – that it must be generated through working with the community over a protracted period of time. In their reflections, students wrote about how they developed admiration and respect for the community leaders who have dedicated many years to these projects. The process of learning outside of the classroom through partnerships opened my eyes to how sustainable practices can be taught in an efficacious manner. The discussions are no longer theoretical when students meet leaders in the community and work with them directly on environmental projects. The students are given the opportunity, if they engage in self-reflection, to understand their privilege and the unequal manner in which environmental degradation affects communities, to witness the power of partnership, and to experience the network of support needed to enact change.

CONCLUSION

The projects I discuss outline a circuitous route. Among the demands of teaching, committees, office hours, grading, preparation, research, and family, it is difficult to enact wholesale change. I own a hybrid vehicle, but I commute long distances. We compost, but we are not vegetarians. I support environmental charities, but I still fly to conferences. It is challenging to fully dedicate all of one's actions to sustainability, but like gardening, the process is not instantaneous, it is slow.

Incremental creative changes in my personal life reverberate into all areas of my faculty position. I believe the most important thing that I can do as an educator is to ask important questions that lead me, my students and colleagues to become critical thinkers and consumers. In order to authentically ask these questions, I test these waters in my own livelihood. I am not dissuaded by the lack of immediate results because the issues are so immense that solutions are not simple. Focusing on personal sustainable actions is an effective way to center oneself and gain foundational knowledge. Envisioning a new sustainable reality locally lays the groundwork for expanded actions in the broader community. The pressures of environmental degradation are intense, the future appears bleak, but I am encouraged by students' increasing level of concern for the environment. If we are going to advance sustainability it will require interdisciplinary cooperation and collaboration. It will also necessitate a bit of creativity and imagination.

NOTE

1. To see videos of the kinetic artworks discussed please visit: http://billyfriebele .com/

REFERENCES

Anacostia Watershed Society. (n.d.). *Our mission.* https://www.anacostiaws.org/ (accessed on January 4, 2021).

Chesapeake Bay Trust. (2021). *Raincheck rebate.* https://cbtrust.org/grants/prince -georges-county-rain-check-rebate/ (accessed on November 15, 2020).

De Certeau, M. (2011). *The practice of everyday life* (3rd edn, trans. S. F. Rendall). Berkeley, CA: University of California Press.

Gutiérrez, L. G. (2017). Rasquachismo, in *Keywords for Latina/o Studies* (pp. 184–187). New York, NY: NYU Press.

Levi-Strauss, C. (1968). *The savage mind.* Chicago, IL: University of Chicago Press.

O'Donnell, R. (n.d.). *Single-use river: Microplastics in the Anacostia River.* https:// www.arcgis.com/apps/Cascade/index.html?appid=0ff0d351069a477c915570 513f01d082 (accessed on November 15, 2020).

Perl, L. (2019). *Community gardening group branches out with urban forest program,* 5 June. https://www.baltimoresun.com/maryland/baltimore-city/ph-ms-forest -network-0604-20150604-story.html (accessed on December 2, 2020).

ReCreative Denver. (n.d.). *ReCreative Denver: Creative refuse store and community art centre.* http://www.recreativedenver.org (accessed on November 23, 2020).

Tate Modern Museum. (n.d.). *Arte povera.* https://www.tate.org.uk/art/art-terms/a/arte -povera (accessed on November 23, 2020).

PART II

Internal/external integration (values to action)

8. Spanning a sustainability career: challenges, changes, and commitment – an interview with Dr. Paul Shrivastava

Shelley F. Mitchell

INTRODUCTION

As academics and scholars working in Sustainability and Management Education, we navigate within institutions of higher education, teaching students, conducting research, and serving on committees, which contributes to our professional lives. In our personal lives, we may engage in our local community, volunteer at a child's school, participate in a clean-up project, or serve on a town committee that contributes to its betterment. All these individual efforts contribute to sustainability and social capital. Often, we compartmentalize our professional and personal pursuits without realizing or appreciating the influence they contribute to the whole person and to our colleagues, students, friends, and family.

As leaders in our classrooms and communities, we have diverse opportunities to lead by example, to have an impact, and to make a difference. Along the way our personal and professional sustainability actions overlap similarly to the three spheres of sustainability: social, environmental, and economic (Elkington, 1997). Kanashiro, Rands, and Starik (2020, p. 824) define personal sustainability actions as "individual efforts that demonstrate care for self and others, and especially the planet and its biosphere, while acknowledging the interconnectedness of all living entities at all levels. By impact, we mean our ability as faculty to model behavior, and to enlist other individuals to engage in, personal sustainability actions that will result in a positive, collective transformation."

Dr. Paul Shrivastava (Figure 8.1) has demonstrated numerous personal sustainability actions and it was an honor to interview him for this chapter. As we are both members of the Academy of Management (AOM) and Organizations and Natural Environment (ONE) Division, I had the pleasure of hearing Paul

Figure 8.1   Photograph of Paul Shrivastava

present many times. One talk was about working towards a sustainable global future by developing a deeper understanding of complex Earth systems and human dynamics across disciplines. His presentation resonated with me because he spoke with passion, commitment, and hope. The same qualities he shared during our interview. At the time of the interview, in his capacity as Chief Sustainability Officer, Paul was working on launching the opening of the Penn State University's fall semester amid the COVID-19 pandemic.

BACKGROUND

Early Life and Education

Paul was born and raised in Bhopal, India, where he attended St. Joseph's Convent School, and Maulana Azad National Institute of Technology earning a bachelor's degree in Mechanical Engineering. From there he received a graduate degree in Management (MBA) from the Indian Institute of Management, Calcutta. After working in management roles, he earned a PhD from the University of Pittsburgh. Paul has a unique background that combines academic scholarship and teaching with significant entrepreneurial and senior management experience; we will learn more about this below.

Academic Career

Paul held the Howard I. Scott Chair in Management, a distinguished professorship at Bucknell University, and was Associate Professor of Management at the Stern School of Business, New York University, where he earned tenure. He was awarded a Fulbright Program Senior Scholar award to study Japanese corporate environmental management at Kyoto University, Japan. He has also taught at the Helsinki School of Economics, the ICN Business School in Nancy, France, and the Indian Institute of Management (IIM) in Shillong, India.

Paul was Distinguished Professor and Director of the David O'Brien Centre for Sustainable Enterprise at Concordia University, Montreal, Canada. He is the author of *Bhopal: Anatomy of a Crisis* (1989), a book that launched the field of organizational crisis management. He co-founded the ONE Interest Group, later to become a Division of the Academy of Management (the world's largest academic professional association in Management Studies). He also founded *Organization and Environment*, a journal of SAGE Publications.

In 2017, Paul was appointed the Chief Sustainability Officer at Penn State University, Director of the Sustainability Institute, and Professor of Organizations at the Smeal School of Business. In October 2018, he became a full member of the Club of Rome, which has served as a source of inspiring work for over fifty years.

As a prolific researcher, Paul has published 18 books, and over 120 articles in refereed scholarly, and professional journals. A highlight of Paul's academic career includes an impressive 173 publications and 15,404 citations (Microsoft Academic, 2021). He has served on the editorial boards of leading management research journals including the *Academy of Management Review*, *Asian Case Research Journal*, *Strategic Management Journal*, *Organization*, *Risk Management*, *Business Strategy and the Environment*, and the *International Journal of Sustainable Strategic Management*. In addition, his work has been featured in the *Los Angeles Times*, *The Philadelphia Inquirer*, *The Christian Science Monitor*, *The Globe and Mail*, and *The Gazette (Montreal)*, and on the MacNeil/Lehrer NewsHour.

Management Career

Prior to working at Penn State University, Paul served as Executive Director of Future Earth, a global research platform for environmental change and transformation to sustainability. He also served as the UNESCO Chair on Arts and Science for Implementing the SDGs, He is an advisor to several organizations, including the Research Institute for Humanity and Nature in Kyoto, Japan, the International Social Sciences Council in Paris, and Future Earth, and he has served on the Board of Trustees of DeSales University in Center Valley, Pennsylvania.

Paul has created several entrepreneurial ventures as he was part of the management team that launched Hindustan Computer Ltd. (one of India's largest computer companies). He founded the non-profit Industrial Crisis Institute, Inc. in New York, and published the *Industrial Crisis Quarterly*. He was Founding President and CEO of eSocrates, Inc., a knowledge management company. Lastly, Paul is an avid fan of Argentine Tango.

INTERVIEW

The following is the transcript of my interview with Paul conducted on July 22, 2020:

As a way of introduction, could you please describe the most salient aspects of your sustainability career, including your current positions? And the general characteristics of the impact you think that you have had on both the academic and practitioner side of the field of sustainability?

The most seminal thing that influenced my own career and thinking goes back to the mid-1980s with the Bhopal, India industrial disaster. Since it was my hometown and I had grown up there, I went and studied it. It made me change my whole intellectual trajectory. Before that I was much more of a corporate strategy or strategic management and policy person, which is what I had done my PhD dissertation on. My dissertation was on organizational learning within a strategic planning context. That disaster showed me the importance of looking at technology, industrial crisis, and disasters as the sort of unintended price that communities and society pay for economic and technological development: what we call progress in modern life. From then on, I changed my research agenda, I changed my teaching program and curriculum towards understanding techno-environmental crises and disasters, what at that time was called "industrial crises" within the broader context of environmental crises.

By the time the Brundtland Commission issued its report in 1989, I was already starting to think about the implications of sustainable development for corporations. Because it was clear to me that corporations were the main engines of wealth creation, and they would need to become main allies and players in addressing the Sustainable Development challenge. The Brundtland Commission had not paid that much attention at that time to corporate sustainability. There was a big intellectual agenda to be filled, and I started working with many members of the Academy of Management [AOM] on articulating that agenda. In the United States, during the 1990s, we started petitioning for creating the Organizations and Natural Environment [ONE] Interest Group for a few years before it became a Division. That was a kind of extended group of intellectual kinship that we created together. It included some struggles, because within the Academy of Management setting, the discussion of environment was within the Social Issues in Management [SIM] Division. And, SIM was not too happy about taking environment out and making it a separate interest group and division. I see all these things as a formative time for me, for thinking through the big questions of what needs to be done to make global corporations and global economy sustainable. There were two conclusions that came out of that period. One was that this is not just an intellectual task, although there were big intellectual questions that needed to be answered, and a lot of research needed to be done. ONE was getting created to do that. But the answers would not come from doing intellectual work alone. We needed to have one foot on the ground, in practice within companies to bring about change in action. Since the mid-1990s, I have pursued a dual trajectory of building the intellectual understanding of corporate sustainability and trying to work with companies, policymakers, community members, and activists to make sure that action takes place on the ground, in the real world.

At different points in my career, I have partnered with multiple stakeholder groups. It is not always corporations; sometimes it is policy or activism groups. Sometimes it is community; sometimes it is with government policymakers or legislators. The idea is that action and intellectual activities need to be brought together periodically to get things resolved on the ground. This has been a very productive approach for me. I find it very satisfying to see practical results and intellectual developments in combination. On the other hand, it has also been frustrating because we live in siloes in academia. Academic institutional mechanisms encourage one kind of activity – academic research and teaching. They do not encourage us to work both in academic and the real world. Thus, there was a lot of frustration early in my career from not being able to make real-world impacts. But as I have matured, I have found that virtually in any system, there is opportunity to make changes.

You need to understand how change occurs in large systems. The fabric of change is complex and tightly interwoven. You need to first fully understand how change happens, and then figure out ways of enabling it. I work at a big complex university, at Penn State University. Before this, I was in an even bigger organization, called Future Earth, which had 50,000 research scientists from many universities doing sustainability research. We had over 100 organizations funding the five global hubs of Future Earth. The National Science Foundations of the G20 nations funded our research and our goal was to figure out how to make change at a planetary scale. Here at Penn State, we have a hundred thousand students, an annual budget of $6.8 billion, which is incidentally larger than the GDP of 40 countries. So, understanding the system is probably the most difficult part of change. I try to figure out how the current system works, then make some modest goals, and then walk the whole organization towards them in a collaborative process. This is how I can address change in a big organization.

How do you personally practice sustainability in the various aspects of your life?
You have discussed the work front, but what about at the play front, on the home
front, and things that you would like to do in the future? To whom do you communi-
cate those sustainability actions for setting an example?

When I started the intellectual work in sustainability, one of the things I did right
in the beginning, was calculate my own ecological footprint. I went to one of the
first calculators and I calculated my own eco-footprint. It was very interesting to
me to realize that before I came to the United States when I was living in India, my
average eco-footprint was 1.8 tonnes of carbon per year. And I was a very successful
executive in a corporation coming out of a very successful top-rated MBA program.
When I arrived in the US as a graduate student on student wages, living near downtown Pittsburgh (Oakland), in what was a student ghetto, my eco-footprint went to
20 tonnes of carbon. Every few years, I do this exercise to test my eco-footprint. By
the time I became a professor with a salary and a house, my eco-footprint rose to 28
tonnes per year. So, my personal engagement has been driven to look at where my
eco-footprint is and see, what can I do? Thus, moving from a suburban house into an
apartment in a town was one of the biggest eco-footprint reduction exercises that we
undertook. When we moved to Montreal, we lived in an apartment building. It was
a big crunch down from the space that we had in a suburban house in Allentown,
Pennsylvania.

On a day-to-day basis, there are lots of things that can make a difference on one's
eco-footprint. For example, food has a big impact on my eco-footprint. I have never
been a complete vegetarian in my life, even when I was growing up in India. I grew
up in an environment where meat was not an everyday food. It was expensive, it
required an effort to procure, and so it was an occasional food. When I came over to
the USA, I could have bacon at breakfast, a meat sandwich at lunch, and then meat
for dinner. It was affordable, and conveniently and readily available. I deliberately
rejected that high-meat food lifestyle and went back to what I was accustomed to.
I still eat meat occasionally, I still occasionally eat fish, but I am very mindful that
it should not exceed a certain number of times during a week.

On the transportation front, I have always commuted by bicycle to my office.
I buy my house or home in a place where that is feasible. I have ridden my bike to
work for 40 years now, including in cities like New York City, and in Allentown
and Lewisburg, PA, which are small towns. But Montreal is a big city. We lived
a couple of miles from work and we had a bicycle path to my office and whether it
was summer or winter, five inches of snow or rain, I would still bike to the office.
And when people see this, it is symbolic, it has impact. Now that I am in a management position, and people see that somebody in his late 60s is willing to commute by
bicycle, they notice it and it changes people's behavior. So, I maintain this practice.
In terms of a car, I do not like to drive, and I do not drive if I can avoid it. Over
the last 40 years, I can count the number of times I drove. I have been lucky that if
I am going on a vacation, my wife will drive because she is happy to drive. For the
last 10 years, we have owned a Prius, and I feel somewhat guilty to say that next
Saturday we are going to be the proud owners of a fully electric vehicle, and it is
a Tesla Model Y [*smile*].

To answer your question, you can reduce eco-footprint in everything that you
do. You have got to be responsible for your karma, whether it is your food, your
transportation, your vacations, and the houses you live in. We wanted to put a solar

roof on our house; we have not yet been able to do it because our roof cannot take panels in the current form.

Regarding, who do I communicate this to? It gets communicated to my immediate family or children, to our friends and others close to us. I also make it a point to raise these issues with my peers. We have a newsletter called *Mainstream at Penn State*, where we started a small three-question column in it, on what are you doing in your personal life to be sustainable. We asked the provost of the university, and several college deans. It's an easy kind of column to do, but it brings to the attention of 17,000 employees that the Provost actually does recycling of his milk canisters, or other leaders do X, Y, or Z because they point out these little things that you would not otherwise imagine. We focus on personal behaviors because they serve as models. I am convinced that unless you model sustainable behaviors you will not see change. You can keep writing any number of academic papers you want, and you will not see change. But if you do things and people watch it and you let them observe your actions and ask you questions, then you are likely to get more change.

What are some of the challenges that you have faced in implementing your own sustainability practices? What are your efforts to communicate those practices to your stakeholders?

An important lesson was understanding why my eco-footprint was so much lower living in India as a successful executive versus when I came to the US and was a poor graduate student. That analysis showed me that the biggest carbon impact was not from my individual decisions. It was from the systems that I was embedded in. The moment I go work in a building my carbon gets connected to the building carbon for eight hours a day. If I go and see a movie or to a restaurant, all that carbon associated with my surroundings is part of my carbon use. I cannot really do much to change that. And that is why bringing it down from 28 tonnes to currently around 17 tonnes has been a journey. Much of the carbon that we are currently committed to is systemic carbon. I have started seeing systems in which we live and work as a challenge, and the need to make systemic changes. Because of this reality I would like to make the 2,000 buildings at Penn State low carbon.

Penn State just did a Power Purchase Agreement for 70 megawatts of solar power from an installation in Franklin County. That is going to decarbonize 25% of the entire Penn State annual energy consumption. That is going to have a bigger impact on my carbon than what I do in my little 2,000 square feet home. We should be looking at the bigger systems and trying to figure out how to change them because we are all part of those larger systems. It is not easy, it takes time. I can get some marginal savings of a few pounds of carbon by doing things around the home, but I can get much bigger results and changes for a lot more people by changing the bigger systems. I will give you another example. When I arrived at Penn State University, I asked our Teachers Insurance and Annuity Association [TIAA] retirement people if there is a carbon neutral investment fund that our employees can invest in, and there was not. They had not thought of this as a way for individuals to move their retirement money to sustainable investments. So just a few months ago, after a whole year of working through the retirement decision making procedures and systems with TIAA, the retirement committee, the labor unions, the pension department, etc., we are now able to make that an option. Now, any employee who has a TIAA retirement account can put some of their savings into a carbon neutral fund. That is going to move funds within a large pool of $6 billion. I cannot imagine any personal investment decision that could be so consequential because my personal investment value is small. This kind of systemic thinking that connects the personal to the systemic and seeks whole systems change is both exciting but also filled with barriers.

What advice can you suggest to upcoming faculty members in their own efforts to increase personal sustainability practices? And how to communicate those to their own stakeholders?

First, I will say that they should be totally unabashed and unapologetic in their sustainability practices. Personal sustainability practices have huge symbolic value and should be openly displayed and shared with colleagues and students. They should lift them up, make them visible, make them transparent, be proud of them, share them in their classrooms, share them outside the classroom, in their professional spaces, share that kind of thinking and values, and practices in the communities that they are part of. Two things can make this effective. One is talking about them, even in an environment which is very polarized, as it is right now in this country. People just tend to keep quiet and slink away and go and do their sustainability thing in private. I do not think that is a useful response. I think we need to bring issues out into the open. We need to engage people in open conversation, and lead by example. We need to do things and let people see and make judgments, they need to know about who you are as a person. I think we need for professionals in our community of academics to not worry too much about how we are viewed. We are all overly concerned about public persona and private persona. And we try to separate those out. But I think the solution is really bringing those two together making your private public around sustainability questions. Do not be apologetic if you do not want to eat a meat dish. Okay, it will be a little inconvenient. Maybe you must make an announcement. And people just shy away from these things, although there are some people who make a big drama out of it too. But by and large, I feel like people are apologetic and I think that is the wrong way to do it. You just need to be open, just do it. And ultimately, people will see value in it.

What is your general assessment of the United Nations Sustainable Development Goals (SDGs) as one means for faculty to upgrade their own personal sustainability agenda? And to what extent do you yourself use the SDGs for these purposes?

The Sustainable Development Goals were created with a vision that is probably the furthest away from personal action. They were created at a global level for national policymakers to move their economy towards sustainability and address the big questions. As a set of values and action areas they make sense individually too, but there must be a translation from the global and national levels to the individual level. Let us just take one or two goals like hunger, or poverty. What can we be doing about hunger or poverty is very different than what a mayor of a city can be doing, or the prime minister of a country can be doing? This translation, unfortunately, has not happened in the last five years since these goals have been articulated. And the project that you and others are working on, is frankly, one of the first public and community-wide projects that I am seeing in this area. Although, I tried to start something like this in a program that I teach in France about three years ago. A colleague and I did a class project about what do SDGs mean personally to you.

But I think this work that you are doing is the first systematic analysis of it. It is very important work. The SDGs can serve as a good guide for personal action if they are translated well. The translation is a very satisfying exercise because it makes you realize what hunger or poverty might really mean. Let me give you an example of the hunger goal. My son went to Africa as a Peace Corps worker four years ago. And this was at the time when the Sustainable Development Goals had just been announced. He was working in Senegal and Liberia for a time. After he was there for about one year, I met him and asked, "What is the biggest problem for you living here?" because Peace Corps workers live within the community they serve. He said, "I am always hungry." I was shocked by that. I asked, "Why are you hungry? I mean, you got $10,000 in your bank account. You are living in a place where there are restaurants where you can go and eat." He said, "No, you do not understand hunger. I am hungry because I live in a community that does not have enough food in my village in northern Senegal, and you cannot eat by yourself. In this community, when you eat, you share. I cannot buy food for the whole village. I just prefer to stay with the kind of food regime that people here live with." That started a conversation about what it means to be hungry. We discussed chronic hunger and homelessness. So, hunger had been interesting to me as a research topic. The discussion with my son, and the experience of traveling in Africa and India, gave me this appreciation that one needs to experience hunger, in its different forms. We need to experience hunger deliberately and purposefully, to understand what it is we do not really have in our world. If we do not have real occasions to experience this, hunger only becomes a concept to us. Even though, it is such a real thing. That is why I have made an effort to reinterpret each of the Sustainable Development Goals in personal experiential terms, whether it's hunger or poverty, in my own way to personalize the goals on a level that I can relate to. It is the starting point to understanding them and taking them to the next level, to the town you live in, or the state you live in, or the country you live in, on every level. I think you must live it and you must embody it to understand it fully and then you are able to raise it up to the community level and participate in food kitchens or your church, wherever you can be helpful.

How can a transdisciplinary approach incorporating the natural sciences, social sciences, and the arts, extend our understanding of sustainability in ways that moti-vate, inspire, and elevate our human spirit in connection to nature? And how has this type of approach impacted you?

To answer that question, I will refer you to a paper that recently came out on trans-disciplinarity in a journal called *One Earth*, which is a new journal by Cell Press [Shrivastava, Stafford Smith, O'Brien and Zsoinai, 2020]. This paper provides my extended answer to the question of why transdisciplinarity is such a big and complex question since it is very hard to encapsulate in a short answer. To me transdisci-plinarity is not interdisciplinarity between Natural Science and Social Science, it is not multidisciplinarity or pluridisciplinarity combining many sciences. Instead, "trans" stands for going beyond disciplines all together. The problem that must be studied does not come from a disciplinary gap. It comes from the real world. Transdisciplinarity seeks to solve problems of the real world and brings together knowledge from different areas. Transdisciplinarity involves co-creation of knowl-edge with stakeholders. The result of transdisciplinary work is not just publishing a paper in a journal or a book but solving the problem. My transdisciplinary scien-tific work benefitted from being at Future Earth. In fact, the formation of Future

Earth was premised on the realization that for at least 50 years, we have known all the sustainability metrics that we are measuring are getting worse. Despite all the traditional science research (in the natural sciences, ocean sciences, biodiversity, and agricultural sciences), things on the ground have become progressively worse. What is it that we are doing in science that we improve understanding of problems, but do not solve them? To me, it became clear that what science was doing is trying to understand problems deeper and deeper in more rigorously metricized and more accurate ways, and they were not charged with creating the solution. The way solutions would emerge, was by throwing the research results over to policymakers or to corporations and saying, now you try to apply this and solve your problems.

That model of science, which has been around for 300 years, is not adequate for solving the challenges of the Anthropocene. Transdisciplinarity to me means picking up real-world problems, co-designing and co-creating knowledge in a unified way that solves problems, and solving the problem collaboratively with stakeholders, and measuring effectiveness by impact. This kind of science will take a few decades to mature. Transdisciplinarity is going to require a change in culture of scientific institutions like universities. It is going to require change in science funding models.

I am here at Penn State, and its research budget is around $1 billion a year. It is a huge scientific enterprise. How do you create $1 billion worth of research in the university? Well, you must incentivize people, you have to tenure them. You must measure what they are doing. All these things are geared to a disciplinary science model. Our university pays great attention to interdisciplinarity as a way of expanding knowledge. We have seven interdisciplinary institutes. I do not think scientists do what transdisciplinarity really requires. It is a huge task. All this is explained better in the paper in which we give a series of suggestions around funding and structuring research, around research cultures and around research questions that need to be answered from a sustainability standpoint.

Recently I have been thinking about the "transdisciplinary person". What does it mean to be breaking disciplines, of standing outside the disciplines, and looking at unified knowledge creation as an individual? I have concluded that we as knowledge producers, are bound by many different personal disciplines. We have a religion, we have some ethnic ways of thinking, we have a linguistic discipline, you might speak English and Spanish or a couple of languages, but there are 5,000 other languages that you do not speak. We have our own ways of creating individual personalities, which are discipline bound. To truly be a transdisciplinary person one needs to be able to break through one's scientific, engineering, economics, linguistic, ethnic, racial, and other trainings. As I age into my seventh decade, I am working on that, trying to become that transdisciplinary person.

In conclusion, do you have a favorite saying or quote or a poem that encapsulates your sustainability journey that you can share with us?

I like to use this one quote and I have used it for 30 years. This is a saying that my mother used to tell us when we were growing up. It's a statement of values and it says something like, "If wealth is lost, nothing is lost. If health is lost, something is lost. If character is lost, everything is lost." I got that drummed into me when I was young, and I drummed it into my kids and my son even tattooed it on his inner arm! You can take that as a kind of general guiding value in terms of my life's priorities. It is the values, the ethics, the character that is probably the most enduring for me.

Are there any other bits of wisdom you can give us?

Wisdom is a big word. And I am not sure this is a wisdom statement. I feel very strongly about human connection to nature. It has been a passion of mine and I see it in my colleagues like you and many others in ONE, we all share this passion. My word of advice to everyone is to embrace this passion because it is not just a legitimate thing to do, it is a survival thing today. This is the kind of passion that is going to get us out of the challenges of climate change, loss of biodiversity, COVID-19, and all these other crises. It is good news for people who are working on sustainability, not just for career reasons, but because this is their passion. We are at a point in history when there is going to be a pivoting from the materialistic and other ways of defining progress and modernity, to a nature connected way of doing it. I have great hope for people and the communities that we are all part of, that we are probably at the right time and at the right place now.

REFERENCES

Elkington, J. (1997). *Cannibals with forks: The triple bottom line of twenty-first-century business*. Mankato, MN: Capstone.

Kanashiro, P., Rands, G., and Starik, M. (2020). Walking the sustainability talk: If not us, who? If not now, when? *Journal of Management Education*, 44(6), 822–851. https://journals.sagepub.com/doi/10.1177/1052562920937423

Microsoft Academic (2021). Paul Shrivastava [search results], at https://academic.microsoft.com/search?q=Paul%20Shrivastava (accessed on September 20, 2020).

Shrivastava, P., Stafford Smith, M., O'Brien, K., and Zsoinai, L. (2020). Transforming sustainability science to generate positive social and environmental change globally. *One Earth*, 2(4), 329–340. https://doi.org/10.1016/j.oneear.2020.04.010

9. Living and communicating personal sustainability

Amy K. Townsend

INTRODUCTION

For many years, I have been personally and professionally committed to sustainability, yet sustainability can be downright elusive. We humans were born into and partake in non-sustainable economic and social systems not of our making so are sometimes faced with imperfect, non-sustainable choices. As a result, determining how to behave sustainably is not always clear or easy.

Turning to sustainability research for clues into living sustainably may be only somewhat helpful. Most sustainability research focuses more on the collective or macro level of sustainability – e.g., energy infrastructure, economic policy, agricultural and building practices – than on the individual level (Parodi and Tamm, 2018). Similarly, the United Nations' 17 Sustainable Development Goals address the collective scales of communities, nation, and globe. Yet sustainable learning and activities at all levels – from the macro to the micro – need to be addressed if there is any hope for true sustainability.

Furthermore, sustainable development theory and practice have tended to focus on the world outside of ourselves without also attending to the world within us. Personal sustainability is not just about what we do but also who and how we are. It is about what we believe, what we value, how we nurture ourselves, and how we interact with the world around us. In order to make sustainability possible, it has been suggested that inner and outer sustainability must be interconnected (Parodi and Tamm, 2018). Otherwise, the outcomes can be haphazard and inconsistent.

For those individuals interested in living more sustainably, there are long laundry lists of things to do – from recycling to eating an organic, plant-based diet to using renewable energy. Those lists of things are generally focused on reducing consumption and pollution through conscientious choices and represent our personal intentions and interactions with the outer world. For many of us, more attention is paid to our outer world of *doing* sustainability and less to the inner world of *being sustainable*. In an unnecessary dichotomy between doing and being, it is a well-known irony that many active sustainability practitioners risk approaching burnout because they do not live personally sustainable lives (Cox, 2011; Gaither, 2020; Weinreb, 2020).

A personally sustainable life can be achieved, in part, by remaining grounded and true to oneself without being pulled off center by the drama of the day. Mindfulness activities that quiet the mind and alleviate stress can be important practices for personal sustainability. For several years, I practiced tai chi and Middle Eastern dance. Now, I practice yoga and workout five days per week. I also find that meditation provides me with insights into myself or how I can do better in one or more areas of my life. It generally prompts me to ask the question – what is it you really want? The answer is rarely a material one.

My personal journey in integrating more sustainable thought and practice into my life has been borne not just from the mental calculation that caring for our life support system – Earth – is a wise thing to do. It has also arisen from my experiences with people and places over the course of my life and the feelings and meanings that I have attributed to those experiences. Those three things – fact, meaning, and feeling – have had a profound impact on what and how I have researched, practiced, and taught.

Communicating sustainability theory to a class or to readers of a book without making any but the most futile attempts at personal practice can ring hollow so I have done my best to try new things, learn from those attempts, and communicate them through articles, books, speaking at conferences, and teaching. Below, I document many of my personal sustainability activities and how they have informed my teaching – both in the classroom and beyond.

DISSONANCE & REALIGNMENT I: SWITCHING TO A PLANT-BASED DIET

Like many children, I always loved animals. I was 17 years old when our beloved dog, a little eight-pound cockapoo, passed away. It was deeply painful and also uncomfortable as I began to consider her life and death through a different lens. I realized that it is our own emotions, cultural taboos, and beliefs that differentiate our meals from our pets. I couldn't imagine eating our dog and was unable to mentally or emotionally separate the thought of a hamburger from my beloved pet. This created a sense of dissonance in me as my omnivorous diet, which I had largely taken for granted, became a source of discomfort and, if I was true to my feelings and values, an activity that I could no longer justify. Try as I might, I could no longer eat meat.

That was more than thirty years ago. From that point forwards, I made an effort to learn about industrial food production and its impacts not only on the animals involved but on human and environmental health. As a result, I gave up dairy (and most of my allergies) and eggs. But what made me never look back from the decision to become vegan wasn't just the knowledge that factory farming was bad for animals, the environment, and human health – it was also the fact that I *felt* it to the core of my being. I had resolved my dissonance and found alignment between a part of my inner terrain and my outer interaction with the world.

DISSONANCE & REALIGNMENT II: SUSTAINABLE DEVELOPMENT

During college, I spent two summers working as an archaeologist in Oregon and California and fell in love with the beauty of nature, dark night skies festooned with stars, the howl of coyotes at dusk, the feel of dry desert wind on my skin, and the languid enjoyment of allowing my eyes to wade slowly across expansive Great Basin and Cascade forest landscapes. It was absolute peace and solitude. At 19 years old, I committed my life's work to sustainability. I was in love – what else could I do? After each summer, I returned home to suburban Maryland perfectly set up for major dissonance #2.

The peace of being outdoors surrounded by nature with little traffic and zero distractions was quickly replaced by a rapid lifestyle filled with crowds of people, little wildlife, a jarring and culturally created/valued harriedness, and night skies filled with so much light pollution that only a few familiar constellations, stars, and planets were visible. Returning home was painful, but I was determined to do something in support of our beautiful planet.

During senior year in college as an anthropology major, I decided to study the interface between humans and the natural environment with an eye toward environmental sustainability. The term "sustainable development" had only recently been coined by the Brundtland Commission and was exciting due to its enormous scope and promise. It represented a global desire to live within our ecological means, an antidote to the bad environmental news around habitat destruction, the loss of biological diversity, chronic pollution, global warming, and other salient issues.

After college, I traveled to India with a childhood friend who wanted to visit family there. The following year, I moved to Seattle and started a master's in marine studies and worked with the only anthropologist in the department to better understand human–environmental relations – the interface between our inner and outer worlds. My master's thesis was a historical view on human attitudes toward India's Sundarbans, one of the world's largest mangrove ecosystems, and how those attitudes have affected the way people have interacted with the region. In other words, how have cultural attitudes toward the Sundarbans ecosystem varied over time, and how have those attitudes affected people's relationship with the place? Have they feared nature and sought to "tame" it? Have they revered nature and sought to ensure its long-term health? Or have they viewed and interacted with the Sundarbans in some other way?

There were little data available on the area, and a visit to the region had been scrapped because the U.S. was in the midst of the first Persian Gulf War. I stumbled across a treasure trove of research papers written on the area. As it turned out, the Smithsonian had hosted a conference on the Sundarbans. The papers had never been published but were sitting in an office at the National Zoo in Washington, D.C. I traveled there to meet the conference host. I offered to type up the papers and put them into a proceedings format and was able to use them to complete my master's research. It turned out that, depending on who lived there at the time, the Sundarbans had been viewed as a dangerous place that needed to be tamed or a rich habitat to be protected. Our individual and collective attitudes toward the environment drive our interactions with our environment. How to change unhealthy attitudes and resulting behaviors toward the natural environment remains the million-dollar question.

Following graduate school, I moved back home to Maryland where I took a job as a research fellow at a large environmental organization. It was my job to support in-house research on non-point source marine pollution in the Caribbean and to help wrap up the first-ever book on sustainable development, *Choosing a Sustainable Future*, authored by an impressive array of ex-EPA directors and others (World Wildlife Fund, 1993).

As meaningful as much of the organization's work was in helping to protect species and their habitat, I couldn't help but notice how normal and non-sustainable the company's in-house behaviors were. For example, the organization partnered with known multi-national corporate polluters and was careful to avoid tackling issues that might upset its donors. I discussed this with my manager at the time, and she got the green light for me to start a green task force within the organization. There were several staff members in house who were interested in sustainability and believed that we should be doing better. Despite our efforts, little was done to move the organization's needle toward sustainability because we did not have support from senior leadership.

After my year-long fellowship was over, I started an environmental research and consulting firm in Maryland and spent the next four years conducting research and writing a book about how best to green a workplace based on my experience at that environmental organization. To my knowledge, *The Smart Office* (Townsend, 1997) was the first how-to book on workplace sustainability. It covered everything from the actual greener construction or renovation of an office building to greener procurement and renewable energy.

One industry at the forefront of the sustainability dialogue was the building industry, and I began attending and speaking at green architecture/building conferences. Over the years, I realized that green offices are important, but if we're looking to ensure ecological sustainability – or, as some suggest, ecological resilience (Gunderson, 2000) – we need to promote not just green building but green business. That launched a decades-long effort to learn what I could about how to green companies, not just their facilities.

As my network of friends and colleagues in the sustainability space was growing, I decided to invite some of them over for dinner and hosted a few dinner parties per year. The field of sustainability was still relatively new, and many of the people I knew were excited to find out what others were learning and doing in their own practices. Most of the attendees were friends or friends of friends, architects, builders, engineers, and others who were incorporating sustainability principles and practices into their work.

One of my guests reached out in 2020 to tell me that those dinners had a profound effect on him and his career. An engineer by trade, he caught the sustainability bug and networked at those events, first learning then later teaching all he could about straw bale, cob building, and other greener building technologies. For several years, until I left Maryland in 2003, I held informal sustainability dinner parties a few times per year where I prepared vegan meals for fellow faculty members and other sustainability practitioners who worked for non-governmental organizations, corporations, and for state and federal governments. Those dinner parties were a great venue for sharing both personal and professional sustainability experiences and to take that information back to our students.

It was during this time that I began serving as Adjunct Professor at both the James Madison University College of Integrated Science and Technology and the George Washington University School of Business. Over the years, I co-taught courses in the U.S. and in Malta on sustainability, provided lectures on what I had learned of workplace greening, spoke on sustainability and green building at conferences, and introduced students to a broad array of sustainability-related topics that I was writing about, which ranged from green hotels to zero-VOC products to rainwater harvesting (Townsend, 2006, 2009).

I also realized that most green business curricula do not teach any environmental courses or facilitate ecological literacy. In response, I designed and taught a course for business students that was intended to bridge environmental science and sustainable business. It contained a lab in which students actually had to go outdoors to identify plants and learn about a particular species and how it contributes to its ecosystem throughout its life cycle. This was a new perspective for many of my students and one that often surprised and delighted them as they began to understand the natural world a bit more.

One summer, I was in Malta teaching a small group of students from James Madison University about sustainability. That summer, my students were interested in developing a plan to collect waste cooking oil to be used in bio-diesel production. As they conducted research and began to outline their plan, I met with the country's environment minister, who became interested in the idea of biodiesel. At the end of the summer term, I arranged for my students to present their findings to the environment minister on the overall feasibility of biodiesel in Malta and their household cooking oil collection plan. This was a great experience for the students, who later co-authored a book with me, *Exploring Sustainable Biodiesel* (Townsend, 2008).

DISSONANCE & MORE DISSONANCE: RABBIT RESCUE AND THE FOOD BUSINESS

Throughout my career in the sustainability field, I have worked toward being more sustainable at home. I already composted, recycled, used energy-efficient light bulbs, used zero-VOC paint, purchased cruelty-free items, and telecommuted whenever possible. However, I knew I could do more. So, in 2005, I started a non-profit rabbit sanctuary at home.

Domesticated rabbits are the third most discarded companion animal after dogs and cats (Daly, 2017). It's not unusual for someone to buy a rabbit because it's cute or get a pet rabbit around Easter time. After a while, many of those animals end up neglected, in shelters, or dumped outdoors under the errant assumption that they instinctively know how to survive.

I had previously adopted companion rabbits, so I was familiar with their needs and behavior. Having recently bought a house, I had plenty of room to provide a safe indoor home. I wasn't aware of any other rabbit rescue group in my geographic area, so I converted a two-car garage into a large room in the house and eventually had enough rabbits to keep me busy. I got them veterinary care, adopted some out to loving homes, and cared for those that weren't adopted out.

On top of a full-time job, I was caring for about twenty rabbits, which took a fair amount of work. I began learning something important about myself and my need for personal sustainability from this experience. In my attempt to meet a need – in this case, caring for homeless rabbits – I created more dissonance in my life than I relieved.

It took some time for me to fully appreciate that lesson. In Fall 2007, with the rabbit rescue in full swing, I started the first nationwide vegan meal delivery service to meet a need for ready-made, plant-based meals. I hired a professional chef and a small team of helpers. Every week, we prepared a unique menu of entrees, soups, and desserts then packed and shipped them across the U.S.

Like the rabbit rescue, I launched this company because I saw a need that fitted with some of my core values. However, as much as I enjoyed providing quality meals to our customers, the work was stressful, not joyful. I prepared the weekly menu, ordered ingredients and packing/shipping supplies, met the delivery trucks, and provided all customer service. My chef, staff, and I cooked the full menu of items one long day each week in a rotation of three different commercial kitchens. Less than a year later, I shuttered the business in the midst of the Great Recession because it had become financially unsustainable.

MEETING SUSTAINABILITY DEMAND

In the intervening years, I found a happy medium between grueling sustainability work and inner balance. I taught, did environmental consulting, and worked with companies in the biofuels and extreme energy efficiency fields. I coached fledgling green businesses through my work with the Small Business Administration's SCORE Association (SCORE, 2021). I have also collaborated with a past student from the George Washington University on energy efficiency and ocean plastics.

Much of this work stopped around the time of the 2016 presidential election. It was clear that sustainability was not a national priority because it had not been internalized at the individual and cultural levels and still remained largely focused on collective task lists. As a result, I shifted my focus more heavily to personal sustainability – both inner and outer – with more time spent in stillness and reflection. My doing and being activities include:

- **Food**: I follow a plant-based (vegan) diet, compost food waste, and purchase organic produce from Misfits Market, a company that delivers a weekly box of vegetables that might otherwise have been discarded for being too large, small, misshapen, over produced, or close to expiry (Misfits Market, 2021). I also planted an organic herb, vegetable, and berry garden and am experimenting with a vegan ketogenic diet and intermittent fasting.
- **Gratitude**: I keep a daily gratitude journal.
- **Activity**: I practice yoga four to five times per week.
- **Cooking**: I use a lid on pots to hold in heat, am happy to eat leftovers, and compost any food scraps.
- **Lights**: I use LED bulbs and turn indoor and outdoor lights off when not in use.
- **Clothes washer**: I purchased an efficient washer, run it on cold water only (avoids using energy to heat water), and generally use the quick wash setting (cuts wash times by more than 50%).
- **Shower and toilets**: I generally take quick showers and use an outdoor solar shower in warmer weather. We also installed ultra-low-flow toilets and shower heads at home.
- **Community**: I have taken in homeless animals (rabbits, dogs, cats); stop to help turtles cross the road; certified our yard as wildlife habitat with the National Wildlife Federation (2021); use the crowd-sourced (iNaturalist,

2021) phone app to map the species that live around our home; transitioned the stray cats to be strictly indoors to protect the lizards, mice, voles, birds, and other critters who enjoy the yard; allow native plants to take over the unmowed yard in support of the native animals that utilize them for food and shelter; remove kudzu and other invasive plants by hand around the property; allow the wild turkey and deer to enjoy the yard and remnants of the garden; have practiced small-scale beekeeping for several years; and seek entertainment by sitting outside listening to the barred owls who spend their twilight hours discussing the day's events.

- **Purchasing**: I use natural/compostable kitty litter, toilet paper made from recycled post-consumer-content waste fiber, and cruelty-free personal care products. Whenever possible, I buy clothes from ThredUp (ThredUp, 2021) and thrift stores and purchase most furniture and appliances from Craigslist.
- **Insulating**: I have added insulation to my home attics.
- **Telecommuting**: For much of my career, I have had the privilege of working from home, which has been environmentally and personally beneficial.
- **Repairing**: I repair appliances and other products whenever possible rather than buying new (learning to repair appliances is not only easy but also saves tremendous amounts of money and is easier on the environment).
- **Education**: I have taken online environmental courses at Harvard to deepen my understanding of the vast field of sustainability and shared my own experiences with the professors and fellow students through online course assignments.

CONNECTING THE DOTS: TYING PERSONAL SUSTAINABILITY PRACTICE TO THE UNITED NATIONS SUSTAINABLE DEVELOPMENT GOALS

Through the years, I have talked with students, faculty, and other professionals about sustainability. Often, there is a tangible frustration that environmental challenges are so enormous that it can be difficult for a single person to make a positive difference. Yet, collectively, what each of us has done *has* made a difference – sometimes negative (e.g., pollution) and sometimes positive (e.g., healing the ozone hole).

Faculty can help students to connect the dots between large environmental issues and their own personal sustainability practices by encouraging them to imagine how their personal practices are making a difference. For example, they can create a simple matrix that indicates which of the United Nations 17 sustainable development goals their personal sustainability activities have likely impacted – even if on a very small and local scale. For example, my work in green building and green business can be seen to have relevance to all 17 of the goals. This is not a fully scientific endeavor, but it does provide a means to see which areas have impacted the most and which are areas for improvement.

FINAL THOUGHTS

I have spent much of my sustainability career engaged in activities that aligned with my values. Those activities that focused largely on "doing" at the expense of my own well-being and that focused on the outer world at the expense of my own inner world (e.g., rabbit rescue, food business) left me feeling unfulfilled and were discontinued because they were not personally sustainable. One of the most important things that I have learned during my decades-long sustainability journey is that a lasting commitment to sustainability is as much an inner journey as it is an outer one. The lack of a mindful, internally focused sustainability practice can jeopardize any lasting outward sustainability commitment. Conversely, stillness, peace, mindfulness, and slowing down can help lead to a more fulfilling and personally sustainable existence.

Faculty members have the opportunity to integrate a vast array of sustainability practices into their lives that align with their environmental values and that feed their souls. They can communicate their personal sustainability experiences with students for richer, more lived lessons in sustainability. This can occur within the classroom via lectures, workshops, and homework assignments. It also can be facilitated through labs and outdoor experiences that enhance ecological knowledge and ecological literacy, through socials, and through encouraging mindfulness practices.

Personal sustainability covers a broad spectrum of doing and being in both our outer and inner landscapes, and students should be made aware of this. If they anticipate carrying out sustainability work over the course of their careers or their lives, those activities need to be personally sustainable as well. Sustainability offers an opportunity to not only focus on carrying out sustainability activities in the world outside ourselves, which sometimes can feel like a weighty task, but also to nurture ourselves in such a way that our sustainability commitments are truly sustainable.

REFERENCES

Cox, L. 2011. *How do we keep going? Activist burnout and personal sustainability in social movements*. Helsinki: National University of Ireland Maynooth. Online at file:///C:/Users/owner/Documents/Amy/02%20-%20PUBLISHING/Book%20Chapters/LC_How_do_we_keep_going.pdf (accessed January 24, 2021).

Daly, N. 2017. "Here's why Easter is bad for bunnies." In *National Geographic*, April 12. Online at https://www.nationalgeographic.com/news/2017/04/rabbits-easter

-animal-welfare-pets-rescue-bunnies/#:~:text=Rabbits%20are%20the%20third %20most,care%20they%20need%2C%20their%20behaviors (accessed January 24, 2021).

Gaither, C. 2020. "Sustainability and the never-ending battle against burnout." In *GreenBiz*, July 20. Online at https://www.greenbiz.com/article/sustainability-and -never-ending-battle-against-burnout (accessed January 24, 2021).

Gunderson, L. 2000. "Ecological resilience: In theory and application." In *Annual Review of Ecology and Systematics*, 31: 425–439. Online at https://doi.org/10.1146/ annurev.ecolsys.31.1.425 (accessed January 24, 2021).

iNaturalist. 2021. Online at https://www.inaturalist.org/ (accessed January 24, 2021).

Misfits Market. 2021. Online at https://www.misfitsmarket.com/ (accessed January 24, 2021).

National Wildlife Federation. 2021. Online at https://www.nwf.org/certify (accessed January 24, 2021).

Parodi, O. & Tamm, K. 2018. "Personal sustainability: Exploring a new field of sustainable development," in *Personal sustainability: Exploring the far side of sustainable development*. New York: Routledge, pp. 1–17.

SCORE. 2021. Online at https://www.score.org/ (accessed January 24, 2021).

ThredUp. 2021. Online at http://www.thredup.com (accessed January 24, 2021).

Townsend, A.K. 1997. *The smart office: Turning your company on its head*. Olney, MD: Gila Press.

Townsend, A.K. 2005. "Business ecology: The future of green business?" in Starik, M. and Sharma, S. (eds), *New horizons in research on sustainable organizations: Emerging ideas, approaches and tools for practitioners and researchers*. Sheffield, UK: Greenleaf Publishing, pp. 187–213.

Townsend, A.K. 2006. *Green business: A five-part model for creating an environmentally responsible company*. Atglen, PA: Schiffer Publishing.

Townsend, A.K. 2007. "Green business: A call for place-based best practices and sustainability indicators," in Cassar, L. F., Conrad, E., and Morse, S. (editors), *Measuring sustainability: Theory and experience from the Mediterranean*. Trieste, Italy: United Nations Industrial Development Organization and the International Centre for Science and High Technology, pp. 71–82.

Townsend, A.K. 2008. *Exploring sustainable biodiesel*. Atglen, PA: Schiffer Publishing.

Townsend, A.K. 2009. *Business ecology*. Atglen, PA: Schiffer Publishing.

Weinreb, E. 2020. "Combatting burnout in sustainability work." In *GreenBiz*, February 19. Online at https://www.greenbiz.com/article/combating-burnout-sustainability -work (accessed January 24, 2021)

World Wildlife Fund, National Commission on the Environment. 1993. *Choosing a sustainable future: The report of the National Commission on the Environment*. Washington, DC: Island Press.

10. Sustainability-oriented management education as personal practice and a "kit" for managers beyond the era of business as usual

Ralph Meima

In 2006, I was hired by Marlboro College in Vermont to serve as founding director of the Marlboro MBA in Managing for Sustainability. By 2012, the program had fallen short of its recruitment goals and was chronically under-enrolled. In June of that year, my position was terminated. By August, I had gone. Within a few more years, the Marlboro MBA was no more. Indeed, as I write, Marlboro College itself – founded in 1946 and memorialized as one of the "Colleges That Change Lives" (Pope 2006:10) – has recently collapsed, closed, and sold its campus.

Living near Marlboro, I pondered this story as it unfolded in the local newspapers with resignation and dismay – resignation because opportunities to save the program, graduate school, and college as a whole by thinking and acting differently had been missed; dismay because it epitomized the fading away of two institutions I value, one long-lived and one quite recent: the small, private liberal arts colleges of New England, and the constellation of innovative sustainability-oriented MBA programs that sprang up as this century dawned. While many examples of both survive, the existence of the remaining colleges and "green" MBA programs is threatened in our new economic and demographic reality.

My notion of personal sustainability practice was interwoven with the second institution – the "green" MBA. While "walking the sustainability talk" might tend to make us think of physical practices, like recycling, composting, gardening, reducing meat consumption, using zero-emissions vehicles, traveling by mass transit, and weatherizing one's house, I submit here the pursuit of sustainability in management education as a major aspect of my "talk-walking," and use this as an opportunity to examine and learn from the insights gained in this pursuit.

The year 2012 saw the culmination of about thirty years of effort on my part to bring a radical, holistic understanding of sustainability into management education in the vain hope, I now fear, that it would be possible to rationally reorganize and manage our society through the transition to sustainability. This chapter will attempt to frame my experiences relative to that vision. While I now believe that the prospects for rationally managed change are dismal, I will nonetheless propose and outline a future-oriented "kit" around which sustainability-oriented management might be based beyond the present era.

* * *

In college in the early 1980s, I discovered the literature of sustainability, appropriate technology, and clean energy through the works of such thinkers as E.F. Schumacher, Gregory Bateson, Arne Næss, Lester Brown, Hazel Henderson, Ernest Callenbach, Fritjof Capra, and Amory Lovins. A different world seemed within grasp – an attractive world of decentralized energy generation, organic food production, human-scale communities, and ecological rescue.

In the mid 1980s, with a conventional corporate career underway, I volunteered in green political organizing. In the late 1980s, while in graduate school, I organized clubs and recruited speakers in order to bring the sustainability perspective into graduate management studies. One high point was a visit I arranged in 1988 by Ben Cohen, co-founder of Ben & Jerry's Ice Cream, to the Wharton School of Business, where he captivated an at-capacity audience with tales of socially responsible business, topped off by ice cream samples. This was before climate change had become politicized, and that summer there was bipartisan concern in the US Congress that led to widely followed hearings on the topic. At Wharton, other students and I advocated for inclusion of climate change in the sustainability perspective we thought should permeate the MBA curriculum.

By the early 1990s, then living in Sweden, I started work on a PhD in corporate environmental management. Ten years later, I had my doctorate, along with years of experience with teaching, training, researching, and writing about environmental management systems, environmental auditing, CSR/ESG certification standards, life cycle analysis (LCA), ecolabels, socially responsible investment (SRI), industrial ecology, and other tools for lightening companies' environmental and social footprints. With Clinton and Gore in the White House for most of the 1990s, this was the decade of the Rio Earth Summit, ISO 14000, the Kyoto Protocol, and global advances in environmental regulation, sustainability-related management practices and standards, and many new networks. It was a heady time for believers in a rational path to sustainability.

The first decade of the 21st century proved challenging for the optimistic rationalist. In the wake of the 9-11 attacks, America's democracy and reputation showed vulnerability. The Iraq War was sold to the West on concocted foundations. The economy swelled up and then popped. Over the course of that decade, certain trends became increasingly apparent, including the ongoing impoverishment of the American middle class, the weakening of civil liberties, and the growing consensus about the severity of climate change, despite its deniers.

I returned to the US in 2002 with my new doctorate to teach management courses at the School for International Training in Vermont. I soon sensed a gap there between espoused values of social justice, peace, international cooperation, and experiential learning, on the one hand, and, on the other, an atmosphere of defensiveness, culture war, and professional self-preservation among faculty and students trying to make sense of profound change in everything from the economics of graduate education to growing signs of deep polarization in America's political life.

It was amidst such circumstances that I accepted – with great excitement – an offer in 2006 to lead the creation of a sustainability-oriented MBA program at Marlboro College Graduate School. From the start, I was encouraged to collaborate with Gifford and Libba Pinchot, founders of sustainable-MBA pioneer Bainbridge Graduate Institute (BGI), to develop a hybrid limited-residency program based on their experience, adapted to the needs and conditions of the Northeast. From having no faculty, curriculum, marketing strategy, or students at the start of 2007, the program grew into the Graduate School's largest program by 2010, with a vibrant community of faculty, students, advisors, and guest speakers who met in person nine times per year for long weekends of in-person classes, seminars, and events.

It was a wonderful experience, and one of the most productive periods in my life. We recruited a highly qualified and committed faculty based all over the Northeast, and a diverse, motivated group of students. The curriculum blended subjects one would find in a conventional MBA with courses on climate change, sustainability, systems thinking, and CSR. Our own version of BGI's "circle" meeting was a feature of each day of the intensive, designed-to-foster program-wide community building. I valued the convergence of my job with what I regarded as my life's mission and personal sustainability practice: the evolution of organizational management toward a sustainable mode.

The MBA program operated for several years past 2012, when I left, until it petered out due to a lack of students and institutional commitment. In hindsight, I believe the program was among other things a victim of the 2008 financial crash; Marlboro College had looked to the program to quickly generate a financial surplus, to offset declining undergraduate enrollment, but it became clear that the MBA also needed long-term investment in order to survive the crisis and grow.

As I parted ways with a program in which I had invested so much of myself, one disturbing observation struck me: the management approach the MBA espoused – imbued with community building, flat organization, transparency, accountability, respect, courageous conversation, inclusion, trust, long-term thinking, and systems thinking – saw little adoption by Marlboro's leaders as its crisis worsened. Instead, I saw reliance on hierarchy, legal transactions, need-to-know, lack of transparency, short-termism, fear, and defensiveness. The two paradigms stood in sharp contrast.

Sadly, recent years have seen the demise of many sustainability-oriented MBA programs that were similar in spirit and curriculum, and launched around the same time, including Antioch, Marylhurst, Green Mountain College, the original Green MBA at New College (Santa Rosa, CA), and BGI itself. I have wondered whether they all folded or were absorbed into other programs because they were ultimately not governed and managed with the values and tools they themselves espoused. (Most of the remaining programs that claim to incorporate sustainability operate within large, conventional institutions that were not founded upon the singular mission of driving a transition to sustainability; see for example the Princeton Review's top-ten ranking of Green MBA Programs.[1])

In early 2014, I had the good fortune of meeting the founders of a renewable energy start-up – Green Lantern Group – and the past seven years have been spent deeply involved in solar project development. This has allowed me to learn the financial, legal, technical, and governmental aspects of energy development "on the ground." It has also provided a new way for my thinking about sustainability practice to evolve.

Successfully developing large-scale solar projects is the focus of my work now. As I interact with different groups of stakeholders, the goal is always the same: find land, transact agreements, study feasibility, manage design, ascertain community and local government consent, obtain permits, and support financing, construction, and long-term operations and maintenance.

Green Lantern is a fairly conventional firm in sustainability terms. Unlike some of our competitors, it is not a B-Corp or employee-owned, for instance. Green Lantern dedicates careful attention to environmental studies in vetting sites, but – while we personally embrace Vermont's ethic of land stewardship and the outdoors – we are mainly concerned with regulatory compliance, a very demanding concern in and of itself. At the same time, we keenly pursue the buildout of as much clean, green, renewable solar capacity as possible, in the belief that, by doing well, we will do good.

Green Lantern has built over 50 MW of solar capacity around Vermont – generated by roughly one hundred arrays in over sixty-five towns – serving schools, hospitals, colleges, municipalities, ski resorts, and more. And I believe that, as we promote distributed solar, respond to criticism, explain its logic, and advocate state legislation that fosters further renewable energy development, we are educating the communities we work in about a more sustainability-informed energy future.

While I have not been in contact with graduate management education lately, in the Green Lantern context I am in regular touch with corporate sustainability and CSR staff, environmental consultants, field ecologists and wildlife experts, state environmental regulators, and environmental lawyers with who I work with permits, land-use planning, and related areas. With different idioms than those of academia, the diverse stakeholders in the renewable energy development space nonetheless work where issues of energy, business, public policy, and sustainability intersect. While little interest is directed at elaborating theories or generating new practices for their own sake, a large array of terms and concepts have been absorbed into the world of practice and are used on a daily basis for pragmatic purposes.

Developing commercial renewable energy capacity involves an ongoing dialog with stakeholders – from the supportive to the antagonistic – about such subjects as renewable versus fossil energy, climate change, carbon emissions reduction, embodied energy, energy returned over energy invested, life cycle analysis, habitat impacts and landscape ecology, aesthetic impacts on viewsheds and scenic values, and – least technical – the political economy of who owns, benefits from, and pays for the current transition underway from fossil to post-fossil energy sources, often couched in terms of environmental justice.

To someone like myself who viewed sustainability-related management education as a purpose in and of itself, with immediate bearing on a rational pathway to a sustainable future, this change in perspective has been interesting and valuable. It has given me a window on the steady diffusion of terms, concepts, and procedures from academia into the sphere of practice, pulled by need and entirely pragmatic in nature.

Seen through this window, the demand for such things does not originate in management education, training, or research with a sustainability agenda. It originates in the pressures and challenges of daily professional life, shaped by real estate and contract law, energy facility permitting, other regulatory and public policy domains, community concerns, technology, and market competition. Management academia serves as a generator and repository of narratives and other meaning-artifacts to be drawn upon as needed by all who can use them.

One area stands out in my mind for special scrutiny: the world of voluntary, certifiable management system standards, product standards, ecolabels, reporting systems, and so forth. This is – or, at least, used to be – an important focus of sustainability-oriented academic research and study. Books were written, courses conducted, and theses and dissertations written about the likes of EMAS, ISO 14000, ISO 26000, other ISO standards, the Global Reporting Initiative (GRI), GRESB, SA 8000, FSC certification, B-Corps, organic food labels – to name some of the better-known ones. Taking cues from quality management, voluntary industry self-regulation emerged in the 1980s as a hope-inspiring new dimension of management that showed promise for rationally tackling the challenges of sustainable development. The codes of conduct and standards that emerged in this wave contained many definitions, checklists, and other information that became incorporated into the emerging practice of corporate environmental management, sustainability management, and the like. My own dissertation contains a study of this process (Meima 2002).

In Green Lantern's industry, some of our competitors and potential/existing customers employ these. Vermont's largest electric utility, Green Mountain Power, is a certified B-Corp and uses the GRI format for corporate reporting. Several of our main competitors are also B-Corps. But such things do not appear to play a role in competitive success in this industry. I may be wrong. I have no reliable evidence either way. Green Lantern may have lost business because a potential customer's confidence in a competitor was strengthened by such credentials. However, this has never been obvious, so I can only speculate, and the conclusion I consistently reach is that business is almost always won or lost due to conventional factors.

This looks and feels like conventional business management to me. The federal and state incentives that give this industry its lift may be partly justified on sustainability grounds but the competitive game they fuel appears to reflect business as usual. What paradigms are shattered? What ways of thinking about business, communities, society, and Earth are transcended? I can't think of any.

On the other hand, the physics and technology at the heart of solar energy are boldly transcendent. Sunshine drives the biosphere. The mentality fostered by working with solar energy encourages new thinking about sustainability. Solar energy consumes no fuel, and – manufacturing and construction processes aside – generates no waste. It is all up-front capital investment, in which there is a steady incentive to improve "nameplate" efficiency – that is, the watts of power that each unit of surface area of a solar panel can capture and convert into electricity.

Leaving aside the product and its constituent technologies, however, managerial and organizational life in my professional world feel conventional. Styles and fashions may change from age to age, and the immediate tools may evolve from typewriters to computers and from ledger paper to digital spreadsheets, but I wonder how fundamentally the development of solar projects in 2020 differs from the development of cellphone towers in 1990, minimarts in 1970, shopping plazas in 1950, or gas stations in 1930?

The essence I am grasping for here involves whether management education, training, and research offer unique pathways to a sustainable future.

Judging from what I have experienced in commercial solar development, rational sustainability-seeking action seems to be embedded in and reflected by the public policy and regulatory process, which in turn stems from public opinion and political processes. In other words, public attitudes and demands change; these are translated into legislation and regulation, and society's sustainability hopefully improves as many actors respond. As asserted before, management education, training, and research may remain a sort of warehouse of available items of meaning, but the application of these items happens when a practitioner extracts something and puts it to use within a conventional paradigmatic framework.

* * *

What does the greening of industry look like? Is it strictly material? Can a "green" practice of management education, training, and research unfold alongside the greening of technology, energy systems, materials, products, buildings, and lifestyles? Should it? Will it? Can its pursuit ever live up to the notion of a genuine sustainability practice, alongside material practices such as walking more and eating less meat? If adopted by managers of all kinds and at all levels as an intrinsic managerial function, can sustainability-infused management contribute instrumentally to the rational transition to a more sustainable future?

Or, is the role of sustainability-oriented management academia to remain as the continual generation of ideas and narratives that, while not instrumental as such, form a sort of larder available to practitioners as sustainability imperatives originating in politics and the wider society insert themselves into management, demanding dialog and answers? I hope this distinction is clear.

In view of our recent experience with climate-change denial and assaults on science, a rational society-wide transition to sustainability seems distant. It does not appear that we will analyze, audit, report, and continuously improve our enterprises to sustainability. Conventional management remains ascendant, with its mixture of traditional hierarchy, reductionism, problematic relations to community and justice, opacity, externalization of risks, reactivity, and short-termism. Despite what we now grasp is at stake in the search for sustainability – and it is massive – how can we possibly achieve a desirable future based on a culture of practice that has been much more successful at delivering unsustainability?

Rational organization and management of our common future may be simply beyond the capability of groups of humans; a fatal flaw in most human organizations may be their inability to transcend what is making our current way of life and business world unsustainable.

Should rational design be abandoned? Is its ultimate achievement an impossibility? If so, what next? As the grip of climate change and related crises tightens, can our supremely human mix of bounded rationality, organizational dysfunction, short-termism, reactiveness, and adhocracy achieve what rational design may not: successful adaptation and mitigation? Or will we eventually see the breakdown of economic organization as the effects of climate change overwhelm our persistently conventional way of managing?

If we abandon rational sustainability-oriented management as a way to implement a sustainable economy and society – falling back instead on command and control regulation combined with conventional management – and resign ourselves to a long process of organizational–economic breakdown as the effects of climate change and other resource and ecological crises are felt, is there still a way to deliberately encourage ultimately sustainable outcomes, and can management education and training play a role in this nonetheless?

Consider the following vision. If a rationally managed transition to sustainability is beyond us, but the effects of climate change, habitat destruction, and resource scarcity still allow for a form of civilized society that includes markets, technological innovation, and formal organization, then management education and training will continue to exist in some form. To get there without a rationally managed transition yet maximize the chances that a transition will be successful, we basically have to assemble a "kit." Such a "kit" of management concepts and practices needs to be suited not to today's world but instead to a hypothetical future environment, and then thrown across the chasm representing the chaos and dislocation to which we appear to first need to adapt and then to survive. What will that "kit" look like? Basic characteristics might include simplicity, efficiency, resilience, and a strict avoidance of practices and technologies that have been shown (in the vast experiment of which we are currently part) to lead toward physical and/or social unsustainability. The "kit" will need to transcend the modalities of rational sustainability-infused management, from which system-wide suboptimization continually emerges, and instead foster the emergence of ecological and social health, of which nature provides many examples we can mimic. In this vision, a sequence of partial collapses, scarcity, crises, and endless improvisation replaces a rationally managed pathway – at no small cost to humans and other species. A viable mode of managing beyond the transition is made available through a deliberate process of design, promotion, and adoption.

This sounds a lot to me like sustainability-seeking management academia's ability to generate non-instrumental ideas and narratives that conventional practitioners are able to freely select, experiment with, and incorporate or discard as they deal with ongoing challenges. This deserves deeper exploration. Perhaps the COVID-19 pandemic can serve as an exploratory context.

Virulent global pandemics would seem to pose a threat to sustainability – as one class of such threats. Let us consider the effects of the COVID-19 pandemic on management education, training, and research. Although this article is being written in the middle of this crisis, and it is therefore difficult to have much perspective, we can make some initial, tentative observations. The pandemic has had an unanticipated and massive effect on almost all aspects of our lives. It has driven workers into isolation with their families at home, enormously expanding telecommuting and the use of digital meeting and collaboration tools. This has caused a powerful wave of learning and new skills acquisition. It has also completely upended norms of scheduling, work dress, and the blending of family with work time. Data security and identity protection have acquired new meanings. Managers have needed to become far more flexible and tolerant. Practicality, openness, and results are taking precedence over tradition, rigid control of information, and superficial performance.

Here, Green Lantern can be revisited as an interesting example. The firm has never had a physical office or other location. We work from our homes, vehicles, and job sites. This was done to keep overhead costs to a minimum and maximize operating leverage. It has kept us geographically flexible in terms of where we seek and develop business. It demanded an openness and high level of trust by the two owners, and steady long-term commitment to this model by employees and close-in consultants. It required the use of exactly the types of online digital tools that have suddenly emerged at a premium in the COVID-19 era, such as Box (virtual document management), GoToMeeting (online meetings), Smartsheet (online collaborative spreadsheets), and conference calling. There are a variety of fixed online conference calls scheduled at the same time each week, giving our work pacing and our communication structure. When COVID-19 hit, little changed in our work routines. In a sense, a managerial "kit" had already been deployed six or seven years ago that prepared Green Lantern for a crisis like COVID-19 (and possibly other kinds of crises as well) without rational planning for this kind of scenario. How could we have known? Now, what can this teach us?

* * *

This chapter would not be complete without addressing the question of my material sustainability practices. My interest in sustainable management praxis and the role of education leads off into conceptual realms, but, working close to the big questions surrounding sustainability for a long time has certainly influenced my understanding of my personal duty as a human being, as well. In our Vermont household, personal sustainability revolves substantially around localizing and "organifying" our sources of food and energy. We garden and keep hens for eggs. In our kitchen, we separate organic waste into what the chickens will eat and everything else. The latter fraction is picked up by our town and used to make compost. We keep all of the leaves that fall in our yard, combine them with garden waste, and make compost for the garden through a two-stage, four-bin setup made of discarded pallets. Organics are precious, and we try to retain and circulate as much as we can.

In general, we try to buy locally produced organic meat and generally buy food that originates in our region. Our philosophy is that favoring the local economy unambiguously benefits us, keeping money circulating nearby and reducing energy consumed in transport.

On the energy side, we estimate that we have reduced heating oil consumption by around 80% since we moved into our house in 2003, mainly by installing a wood-pellet stove that heats the core of the house in the winter so we can refrain from burning oil; the pellets come from logging waste generated in Northern New England. This transition also necessitated increased reliance on electric space heaters in more isolated parts of our house, which in turn increased our electric consumption. A significant portion of our electric consumption is offset by a solar net metering group we belong to. Beyond that, we are inspired by the fact that Vermont ranks among the top states in terms of lowest per capita carbon emissions and highest penetration of renewable energy.

We have done some home weatherization, but much more is needed. We wear a lot of fleece and tolerate a house in the winter that is often colder than we'd like. We still drive conventional gasoline-powered cars, but work from home or nearby and thankfully can avoid the worst of American suburban commuter culture. We recycle assiduously but suspect that more is landfilled than we know or want. Our household is far from "sustainable" in ecological terms, but we are trying. It's not practical to analyze everything, and we can't afford to upgrade everything rapidly. The "science" of our household sustainability praxis boils down to, *ceteris paribus*, local is better than distant, organic is better than conventional, renewable is better than fossil, and less energy is better than more energy.

* * *

From my academic focus early on, to my current solar industry work, with personal sustainability practices evolving with changes in life's seasons, it is natural to wonder about what the future holds. My approach to that has been to write and publish speculative fiction about the near future. This provides a way to envision how current conditions and trends might create plausible futures. I became interested in scenario planning as a strategic management tool while teaching strategy back in the early 2000s. This led to attempts to write my own speculative fiction. Following a decade of experimentation and online publication of early drafts, in 2015 the first volume of my *Inter States* trilogy was published: *Fossil Nation* (Meima 2015). The next year, *Emergent Disorder* (Meima 2016) followed. In early 2021, the third and final volume, *Oligarchs' Gambit* (Meima 2021) is scheduled to be published. Neither dystopian nor utopian, the plot (set in 2040) is driven by plausibility and path-dependency, with heavy doses of politics, drama, and (I believe) endearing characters. I will not spoil the suspense, if you plan to read them, but will simply say that – between climate change, fossil-energy depletion, information technology, domestic politics, oligarchy, poverty, and geopolitics – we have certainly created a gripping future to contemplate and prepare for.

NOTE

1. At https://www.princetonreview.com/business-school-rankings?rankings=best -green-mba (accessed May 27, 2021).

REFERENCES

Meima, R. (2002). *Corporate Environmental Management: Managing (in) a New Practice Area*. PhD dissertation, Lund University.
Meima, R. (2015). *Fossil Nation* (Book 1 of the *Inter States* trilogy). Danville, IL: Founders House Publishing.
Meima, R. (2016). *Emergent Disorder* (Book 2 of the *Inter States* trilogy). Danville, IL: Founders House Publishing.
Meima, R. (2021). *Oligarchs' Gambit* (Book 3 of the *Inter States* trilogy). Danville, IL: Founders House Publishing.
Pope, L. (2006). *Colleges That Change Lives: 40 colleges that will change the way you think about colleges* (3rd edn). London: Penguin Books.

11. Learning to think like a city: connecting civic activism with the classroom and the curriculum

Bruce Paton

INTRODUCTION

For me, the journey toward sustainability began with a love for the outdoors, distant from "civilization". The family's vacation cabin and pond in Southern Vermont are still the center of the quiet world. Over the years, my love for quiet, outdoor places has been joined by an intense focus on the built places that encompass our daily lives. This focus on the built environment has intensified in the past few years as my sustainability focus has turned primarily to climate change.

In the past decade, we have seen conferences and publications relating to Aldo Leopold's essay on "thinking like a mountain" (Flader, 1974; Leopold, 1949). Leopold's eloquent, pioneering vision calls us to a deep understanding beyond a simple aesthetic appreciation for Earth's wild places.

The search for sustainability is no longer primarily about figuring out how "natural" places work and how to protect them. Instead, our collective focus is shifting toward understanding the inhabited places and the challenge of making those sustainable. Borrowing from Aldo Leopold, I have come to think of that voyage of discovery as learning to "think like a city". (While many of the problems require regional coordination, the primary lens through which civic actors see decisions focuses on cities.)

Throughout my career, I have interrogated the connection between activism and learning. Two recent phases of that inquiry have deeply affected my teaching and thinking about sustainability. First, I served for eight years, as a sustainability commissioner for the city of Sunnyvale, California, in the heart of Silicon Valley (I served as chair for six years). In that time, our commission passed a climate action plan and Sunnyvale has helped launch a "community choice energy" program that now provides fossil free energy to most of Silicon Valley (Sunnyvale, 2014). In 2019, our city adopted its Climate Action Plan

2.0, with aggressive Green House Gas emissions reduction targets for 2030 and 2050 (Sunnyvale, 2019).

That process has taught me a great deal about how cities make decisions and how they can send out ripples to the communities around them. But cities and city officials think in budget and policy cycles that bear little connection to changes in the physical world. And ironically, the rules for open government inhibit the process of candid and open discussion among the stakeholders in complex problems such as climate change.

More recently I joined the board of Leadership Sunnyvale, a community leadership organization. (I currently serve as Board Chair.) That program identifies and develops community leaders to help drive change in a number of arenas including sustainability. My service with Leadership Sunnyvale has led me to reflect on the deep connections between our region's leading problems – housing, social justice, transportation and climate change. It has also reinforced my impression that the rules promoting open government inhibit the kinds of candid, creative dialogues that will be necessary for real progress toward sustainability.

I have chosen to tell my story by organizing around the UN Sustainable Development Goals (SDGs). This chapter focuses on three aspects of sustainability – environment, social well-being and education – that closely align with several of the UN SDGs.

ENVIRONMENT

One of my longest-standing engagements with the UN SDGs has been focused on SDG 17: Partnerships for the Goals. Following my PhD research, I became involved with a public–private partnership that eventually became a non-profit organization called Sustainable Silicon Valley. The original program was a bold experiment to create an Environmental Management System (EMS) for Silicon Valley as a region. This process involved several large community engagements to identify and prioritize key environmental "pressures" and to pilot cooperative strategies to address them. In 2004, the early stages of that process identified climate change as the highest priority pressure (Melhus and Paton, 2012). The decision to focus on climate change led the organization over time to abandon the goal to create a more comprehensive Environmental Management System (EMS) for the region.

Sustainable Silicon Valley engaged members of the organization to reach out to corporations, city governments and college campuses. Participants created the first voluntary greenhouse gas reduction program in Silicon Valley. Major corporations such as Intel, Hewlett-Packard and Sun Microsystems, joined city governments to commit to reducing greenhouse gas emissions by 20% below 1990 levels by 2010.

The program illustrated the double-edged sword I had been exploring in my research on voluntary initiatives. Voluntary initiatives have the potential to achieve significant advances, but often fail to achieve their targets (Melhus and Paton, 2012). Sustainable Silicon Valley's voluntary greenhouse gas reduction program did not achieve its quantitative targets. But it did help set in motion very extensive climate action in Silicon Valley that continues today. In particular, it led each city to develop a Greenhouse Gas Inventory, years before that became a legal requirement. In addition, State Legislators later credited this industry–government cooperation with helping create the impetus for Assembly Bill 32, California's first public policy committed to specific reductions of greenhouse gas emissions.

Focusing on sustainability in cities addresses a surprising number of the UN SDGs. Specifically, working on Sunnyvale's Sustainability Commission has involved learning to engage with seven of the seventeen UN SDGs from the perspective of a city.

Climate action (SDG 13) has been a major focal point and I have enjoyed playing an active role in the community's efforts to chart a path to carbon neutrality. I am excited that we have been able to chart a credible path to carbon neutrality by 2050, a goal that seemed unimaginable only a few years ago. Now the path seems quite clear, even if some of the issues to be managed do not yet have clear strategies. For example, vehicle traffic now accounts for more than 50% of greenhouse gas emissions, but Sunnyvale and Silicon Valley do not have clear approaches for achieving the established targets.

The one most significant success to date focused on UN SDG 7, affordable and clean energy. The first climate action plan for the city of Sunnyvale was adopted in 2014, during my second year on the Sustainability Commission (Sunnyvale, 2014). The initial plan was comprehensive, aggressive for its time, and almost impenetrable for those that did not participate in developing it. Many citizens and elected officials could name perhaps two or three out of 156 action items in the plan. The city allocated funds to address several of the most easily accomplished items and made vague plans to address the others. Fortunately, the original plan included one transformative idea that captured the imagination of several council members.

The plan called for the city to work with neighboring cities to create a Community Choice Aggregation (CCA) program. CCAs provide an alternative to purchasing electricity from investor-owned utilities, which often have strong incentives to continue providing fossil-fuel-generated electricity. Sunnyvale led the way for all of the cities to join together to create a public–private organization called Silicon Valley Clean Energy (SVCE), to purchase clean energy and distribute it through the existing investor-owned utility's distribution system. Despite political resistance, the city banded together with

11 other cities to create SVCE (SVCE, 2020) (San Francisco, San Jose and the adjoining county of San Mateo have since formed their own CCAs.)

The enabling ordinances for each city enrolled existing utility customers in the CCA, unless an individual customer specifically opted out. More than 98% of all customers in the city switched to the CCA. As a direct result, the city experienced a 25% drop in GHG emissions in a single year and achieved its 2020 goals ahead of schedule (Sunnyvale, 2020).

Silicon Valley Clean Energy has continued to be a driving force for additional steps to address climate change. In particular, SVCE created model ordinances and helped cities develop "reach codes" that committed cities to go beyond State of California building codes. In December 2020, Sunnyvale and San Jose adopted city ordinances to ban natural gas in new construction. The reach codes require virtually all new construction to install electric HVAC, water heating and cooking appliances.

Through my experience on the sustainability commission, my family and I learned quite a bit about energy efficiency and climate action at the personal level. Because of a personal commitment to "walk our talk", during a recent remodel we converted our house to all electric. This required installing a heat pump water heater, a heat pump heating and cooling system, and an induction cook top in the kitchen. At the same time, we installed a charger for my electric vehicle.

The experience has helped me understand the major hurdles that effective retrofitting of residential and commercial/industrial buildings will face in our path to climate neutrality. (I have explored some of those barriers and the lessons learned with my classes.) My family had the means and the motivation to ignore the strong advice from contractors against each of those changes. We received partial rebates for the water heater and the car charger but paid a premium to install the heating/cooling system and the cooktop. We suspect that relatively few other families will have the ability or the motivation to make those changes without strong incentives. As a result, retrofitting buildings is one of the climate change strategies that faces the steepest climb. For a recent "commissioner presentation", I reviewed programs in several leading cities in the San Francisco Bay area and found that none of the cities yet had programs in place to retrofit existing homes and commercial/industrial buildings, despite their significant contributions to greenhouse gas emissions.

As I discuss below, a key part of my efforts to walk the talk has been bringing my learning about the city back to my classroom. My steep learning curve in trying to think like a city has helped me try to model a process of inquiry as the basis for teaching about sustainable business. Business courses follow a dominant logic that focuses on effective management practice and the implications for business performance. Focusing business decision making on the impacts outside the business has been an ongoing challenge for me and for colleagues teaching sustainable business.

SOCIAL WELL-BEING

I have also focused on several aspects of social well-being in my teaching. I have taught courses on business and global poverty alleviation for nearly 15 years at three different schools. When I began those courses the subject was clearly focused on international non-governmental innovations such as micro-credit and Grameen Bank. Although poverty was certainly present in areas such as California, the most interesting organizational innovations were visible in India, Bangladesh, Kenya and other emerging economies.

To my surprise and delight, students in two MBA programs, as well as undergraduates in two very mainstream business schools, have brought considerable energy and creativity to design projects requiring them to devise a business response to a specific poverty problem. In recent years, I have tailored final project assignments to focus on addressing poverty issues in the San Francisco Bay area. Students have responded with creative projects, for example one designed to integrate services for homeless people living in cars and motor homes in a nearby community.

My engagement with the local face of social well-being issues has been strengthened through my wife's work as head of a local non-profit, Sunnyvale Community Services, which serves as the community's safety net organization, working to prevent hunger and homelessness. I have gained many unexpected lessons about resilience through the lens of her work.

A key insight has been the heightened vulnerability of the poor, particularly in a wealthy area like Silicon Valley. Silicon Valley residents are often shocked to learn how pervasive poverty is in an area of such significant wealth.

One particularly insightful learning experience for me has been participating in an annual Poverty Simulation conducted by Downtown Streets Team, Sunnyvale Community Services and a local church. The exercise has participants simulate four weeks in the life of a family living in poverty. Participants go through the exercise of paying rent, food, utilities and other fees, while balancing child care and getting to work, with navigating city and social services. The exercise is an emotional experience because, in a very short time, it simulates the stress that a working family living in poverty would experience. The simulation allows participants to experience the frustrations of not being able to meet their obligations to family members, employers and schools, and simultaneously to feel the humiliation of being denied the respect of their community.

Following the simulation, participants engage in discussions with members of Downtown Streets, who themselves are currently or recently homeless.

Participants in the exercise often express surprise at the skill and courage required for a poor family to manage in the tense environment they have experienced.

The first time I participated, I ended up as a "family member" with the mayor of Sunnyvale. Our family was evicted during one round of the exercise. The experience was so visceral, I heard the mayor describing the experience to his City Council colleagues eight months later. The simulation transformed poverty from an abstract concept to an emotional experience in a period of a few hours.

Leadership Sunnyvale has incorporated the simulation into its training for community leaders. The board and I have worked with the Executive Director for Leadership Sunnyvale to build discussions and activities focused on equity into our year-long curriculum for emerging community leaders.

I was also able to bring a few students from Menlo College to the most recent (pre-pandemic) Poverty Simulation. Students had the experience of working with homeless community members and participants from industry, city government, community organizations. We are exploring possibilities for hosting a poverty simulation at the college when public health conditions permit.

One incident in the community changed my thinking about resilience. Sunnyvale Community Services became the lead organization in coordinating the city's response to a local disaster. A poorly maintained apartment complex caught fire and displaced more than sixty low-income families in April 2016 (KGO News, 2016). Although fortunately no one was killed, the city had never experienced a disaster on this scale in its history. The city's response was superb and illustrated a deep well of generosity. But the effort and cost to relocate over sixty families illustrated the much greater threat to resilience in a community of more than 150,000 in an area known for earthquake risk, and more recently, exposure to risk of wildfire.

The apartment fire was a learning experience for the whole city, a discovery that the usual city and county services would be inadequate in the face of larger disturbances such as a major fire or earthquake. The current pandemic has understandably diverted attention from these natural risks. But the experience has provided a memorable opening to focus City Council members on the need to develop a broader framework for resilience. The experience has also revealed a need for much stronger focus on prevention of disasters like this.

EDUCATION

My experiences in learning to think like a city have contributed directly to my teaching on sustainable business at four different business schools over the past 18 years. I frequently design exercises based on real-world examples I have seen through my work on the Sustainability Commission.

In one instance, I drew on my learning about my city's green building program. I took students to our campus library and challenged them to design a new library on the same footprint. The requirements included making the new building qualify for at least LEED Gold level. I pointed students toward the LEED standards and told them to ask me for help when they needed it. The students rose to the creative challenge and developed a design that appeared to meet the LEED Platinum standards (LEED's highest level; USGBC, 2020).

One aspect of "learning to think like a city" means coming to terms with the fact that we are not educating leaders for the complexity of the problems they need to address. As Aldo Leopold pointed out, thinking like a mountain means thinking about cycles and processes on different time scales and different landscapes (Leopold, 1949). To accelerate progress toward sustainability we need business leaders and community leaders that can think and act on different scales and different landscapes. Thinking like a city also means recognizing that the systems and problems don't respect city, county or regional boundaries.

My focus on changing the conversation about management education has led me on a progression from department chair to MBA Director (twice), to Dean (twice) and more recently back to full-time teaching. Two of those roles have occurred in the context of schools of Business and Public Policy. Those experiences as an educational leader, and my experiences as a civic activist, have strengthened my personal conviction that business education needs to change dramatically to meet 21st-century challenges.

At one of the schools, I taught a series of weekend intensive workshops for nearly a decade. The workshops typically included students from the school's MBA program and from three different public policy graduate programs. On several occasions, I observed significant differences in the skills and tools that the business students and the public policy students brought to the table. Public policy students were much more effective at specifying goals and objectives, defining desired impacts and articulating theories of change. Business students were much more effective at defining business models, developing implementation plans and evaluating financial implications. I came to realize that effective decision makers in an urban setting need a mix of these skills to be truly effective. I have worked ever since to bring public policy-related skills into the business school classroom.

One discovery has enhanced both my teaching and my civic activism. I have realized that students and colleagues embrace challenges such as climate change with much more energy if we engage their creativity and imagination. In one instance, my students had had a very low-energy discussion on the sobering news about climate change from a UN report and a US report

(Davenport and Pierre-Louis, 2018; Mooney and Dennis, 2018). In the next class I asked students to clear their work tables and gave each team a flip chart sheet of paper. I asked them to design a "post carbon" city (Plastrik and Cleveland, 2018). After some initial puzzlement students engaged in very energetic design processes. One of the teams designed a set of changes to traffic in San Francisco to create a more people-friendly downtown. To my surprise, San Francisco adopted a very similar set of changes within two years.

We have lived through an era of profound changes, and the pace of change is accelerating. I draw real hope from watching the instances of real innovation in the city where I live, and in many leading cities around the world. With national governments failing to address the many challenges of climate change, leading cities have begun to demonstrate that change is possible when citizens, governments and businesses choose to cooperate.

As business educators we need to learn from the successes in cities and businesses that have learned to see opportunities in complex problems. For me, walking the talk has come to mean using the finite time left in my career to drive curriculum changes we need to be able to address complex problems.

We must help students learn to understand the challenges we face. But we must help them change their sense of identity from competent business people to business professionals with responsibilities and opportunities to take ownership of community problems. In short, we need to teach them to think like a city.

We desperately need a new kind of leader who can work seamlessly across government, business and community organizations. As a business educator, I have focused on changing the way we educate leaders to prepare them for problems that are qualitatively distinct from the problems our business school curricula are designed to address.

The evidence that this approach to education is having a positive impact is limited so far, but promising. I have been delighted to encounter former students throughout Northern California and see their work in industries ranging from electric vehicles to food processing to financial services. While few would describe themselves as "thinking like cities", many have described how important their interaction with stakeholders has been to their ability to advance their sustainability goals in their work.

As Roger Martin has reminded us in *The Design of Business* (Martin, 2009), the key challenges for business leaders in the next decades are not how to optimize production of complex products, or even how to market products in a globalized economy (although those problems are certainly complex). The key challenges are not even how to lead complex organizations. The real challenge for business education is how to prepare business professionals to contribute to solving complex problems in which business is only one of the key actors.

We need to develop business leaders who are scientifically literate, understand and respect community processes, and are skilled in working with social change mechanisms. That challenge is doubly complex given the need for those people to develop business skills and win the respect of their business colleagues.

For me, learning to "walk the talk" in sustainability has been less about changes in my personal consumption, although my family has significantly changed the carbon footprint of our house, our cars and our diet. Walking the talk has meant advocating for change in my city on precisely the issues I teach about in the classroom. Sadly, walking the talk has often required ignoring the signals about what business schools value and actively reward.

Learning to think like a city has made me understand that the key challenges in achieving sustainability focus on the ways we understand change, how we help others make sense of and act on that change, and how we develop a sense of urgency in both current and future leaders.

REFERENCES

Davenport, C. and Pierre-Louis, K. (2018). U.S. Climate report warns of damaged environment and shrinking economy. *New York Times*, November 23, https://www.nytimes.com/2018/11/23/climate/us-climate-report.html, accessed 1/24/2021.

Flader, S. (1974). *Thinking like a mountain: Aldo Leopold and the evolution of an ecological attitude toward deer, wolves and forests.* Madison, WI: University of Wisconsin Press.

KGO News (2016). Over 140 displaced after fire at Sunnyvale's Twin Pines Manor. April 20, https://www.yahoo.com/news/over-140-displaced-fire-sunnyvales-062523954.html, accessed 9/30/2020.

Leopold, A. (1949). *A Sand County Almanac and sketches here and there.* Oxford, UK: Oxford University Press.

Martin, R. (2009). *The design of business: Why design thinking is the next competitive advantage.* Cambridge, MA: Harvard Business Review Press.

Melhus, P. and Paton, B. (2012). The paradox of multi-stakeholder collaboration: Insights from Sustainable Silicon Valley's Regional CO_2 Emissions Reduction Program, *Journal of Environmental Sustainability*, 2, 29–44.

Mooney, C. and Dennis, B. (2018). The world has just over a decade to get climate change under control, U.N. scientists say. *Washington Post*, http://washingtonpost.com/energy-environment/2018/10/08/world-has-only-years-get-climate-change-under-control-unscientists-say, accessed 10/08/2018.

Plastrik, P. and Cleveland, J. (2018). *Life after carbon: The next global transformation of cities.* Washington, DC: Island Press.

San Francisco Chronicle (2020). San Francisco's Market Street is closing to car traffic: What you need to know. January 21, https://www.sfgate.com/driving/article/Market-Street-car-closure-ban-San-Francisco-15007849.php, accessed 9/25/2020.

Sunnyvale (2014). Climate Action Plan: City of Sunnyvale. Sunnyvale, CA, https://sunnyvale.ca.gov/civicax/filebank/blobdload.aspx?BlobID=23736, accessed 1/24/2021.

Sunnyvale (2019). Climate Action Playbook: Sunnyvale, CA, https://sunnyvale.ca.gov/
civicax/filebank/blobdload.aspx?t=73319.64&BlobID=26529, accessed 1/20/2021.
Sunnyvale (2020). Community Greenhouse Gas Emissions: 2017 & 2018 Update,
June, City of Sunnyvale. Sunnyvale, CA, https://sunnyvale.ca.gov/civicax/filebank/
blobdload.aspx?BlobID=27162, accessed 1/24/2021.
SVCE (2020). Silicon Valley Clean Energy, https://www.svcleanenergy.org, accessed
9/30/2020.
USGBC (2020). LEED v4 Rating System Selection Guidance, https://www.usgbc.org/
leed-tools/rating-system-selection-guidance, accessed 8/11/2021.

12. What do you value? How valuing time leads to deeper environmental engagement

Thomas E. Stone

I.   INTRODUCTION

Timothy Morton classifies climate change[1] as a *hyperobject* due to its complexity in both time and space (Morton, 2013). Sustainability attempts to encompass humanity's response to climate change and must then itself be a hyperobject in order to be relevant. To date, this response has been inadequate, and the concentration of atmospheric greenhouse gases continues to increase; drastically reducing emissions while drawing down those greenhouse gases is the action needed (Jensen & McBay, 2009; Hawken, 2018). The difficulty lies in *how* to accomplish drawdown and reduction. Goal 13 of the United Nations' 17 Sustainable Development Goals (SDGs) directly calls for climate action, but the others (no poverty, quality education, gender equality, affordable and clean energy, etc.) link to our climate emergency response as well (United Nations, n.d.). In fact, the SDGs demonstrate why calling climate change a hyperobject is appropriate. These goals intertwine with each other (and with other environmental issues) in deep, challenging, and sometimes not-obvious or even knowable ways. SDGs provide guidance for sustainable choices, from the personal through global levels. Most of us are unable to directly make or influence policy at the regional, national, or state levels. However, all of us make innumerable "small" decisions daily that are either rooted in environmental stewardship and grapple with these hyperobject-sized issues in the best way we can, or are not. Countless internet lists, blogs, podcasts, books, and other media provide ideas for personal sustainability choices, while sources such as journal articles, case studies, conferences, and professional associations provide similar ideas for institutions. Likewise, there are numerous content choices for university courses; service-learning projects and other experiential opportunities enhance the textbook material that forms the core of most sustainability-related academic learning. Many faculty members likely weave some of their own personal sustainability choices into courses they teach (Coplan, 2019), but Kanashiro et al. (2020) have identified a need for including more broadly this type of andragogy in university coursework.

Acknowledging that "we do not know the right answers, [and] we don't necessarily even know the right questions" (Eisenstein, 2018a), I will nonetheless offer preliminary answers to the broad question, *what should we be teaching our university students about sustainability?*, or more pointedly, *what sustainability behaviors from my own life do I model for my students?* First and foremost, I believe that teaching students to value their time is the most important sustainability behavior that we as faculty can model for our students. Second, I urge faculty to place themselves in situations where they are not the subject-matter experts, in order to model their own learning processes for students' benefit. Both of these actions challenge the traditional faculty role and may require faculty to rethink their classroom approaches; incorporating these ideas into our roles as advisors will be easier. Finally, faculty can choose to incorporate personal sustainability examples by directly linking them to course outcomes.

Clearly, faculty modeling sustainability behaviors that they teach will always resonate more deeply with students, who can quickly discern authenticity from posturing. Kanashiro et al. (2020; see also the references therein) also find that environmental education is enhanced by faculty modeling sustainability behaviors. While the authors mention certain unsustainable behaviors on their part (which they presumably also discussed with students), I would go further and claim that both successes and failures are equally important to share. Elucidating root causes, implications, and ways to avoid these failures is crucial for developing one's environmental understanding and decision-making. For example, the "failure" the authors note related to not being vegetarian only hints at a very deep and nuanced conversation revolving around industrial agriculture, regenerative agriculture, nutrition, social customs, and environmental considerations—this "failure" alone provides a semester's worth of learning.

In this chapter I provide some of my own sustainability choices that I have included in various university courses I teach (though no course contains them all). Connecting course content to these ideas can be a tricky proposition, especially without virtue signaling or students perceiving you as judgmental or overly authoritative; some days I have succeeded and some days I certainly have not. Placing these behaviors within the theoretical lens provided by the Deep Adaptation framework (Bendell, 2020) puts me on solid ground when the going gets murky with my students, and so I will do that here. Briefly, the Deep Adaptation framework asks us, when choosing how to respond to climate change, to consider actions through the lens of *relinquishment, restoration, resilience, and reconciliation.* I find this framework more fruitful than the traditional triple bottom line approach because it challenges us to directly confront relinquishment—what should we go without? The triple bottom line approach reminds us to analyze actions from multiple perspectives, but often-times the hard fact that we need to give up certain things we're accustomed to goes unacknowledged in this framework.

II.   TEACHING STUDENTS TO VALUE THEIR TIME IN ORDER TO MORE DEEPLY ENGAGE WITH ENVIRONMENTAL ISSUES

How can we expect student citizens to make thoughtful decisions with respect to environmental stewardship and sustainability if they utterly consume their post-graduation days with work, the pursuit of traditional consumerist goals, and simply trying to stay afloat amidst student loan debt and other stressors? *I propose that learning to truly value one's time is the most important action related to sustainability that we can instill in our students.* Most universities require that new students take some form of a first-year experience course. These courses cover topics like financial literacy, responsible alcohol choices, various campus resources, and, often, time management. While this is useful, where in the university do students learn to value and respect their time, not just manage it? In what course do students learn that alternatives to the work-to-consume-to-work feedback loop exist, that more does not necessarily equal better, and that you barter for material goods with your time and those goods (with a lot of environmental consequences) will likely not provide any lasting fulfillment? There are certainly courses that challenge and engage students with these questions, but I suspect these are the exception and not the norm. If we are to take the above proposition seriously, we need to model this behavior, while also coming together as a community of educators to deter-mine best practices for this type of teaching.

Responding in the most thoughtful way that I can to questions of environmental stewardship is only possible because I have the necessary time. On campus, I have developed novel sustainability courses, chaired the campus sustainability committee, organized the campus garden and taught Introduction to Gardening, and written about sustainability issues. My formal academic training prepared me for none of these examples, but I have pursued them nonetheless because I have had the time to do so. At home, I grow and preserve a substantial amount of food, keep a small flock of chickens, install energy-saving appliances when necessary and fix others in order to avoid the lifecycle costs of new products, strive to be more of a producer and less of a consumer, and generally live close to the land because I have the time necessary to do so.[2] It takes time to formulate, and then refine, even the questions related to living sustainably. It takes time to learn, to listen, to try, to sometimes fail, and then to try again. This is why it is so crucial that we at least try to help our students value their time, if we want them to deeply engage in environmental stewardship; deep engagement requires that we 'take the time that needs to be taken' (MacFarlane, 2019, pp. 262–263).

Of course, I begin by respecting my students' time (such as employing challenging, thought-provoking work—not busy work, block scheduling courses for my advisees, and truly being present and undistracted when interacting with them). I then attempt to demonstrate how I respect my own time, including these examples:

1. Walking away from a lucrative job that demanded a *total work* mentality and thus choosing time over money (Taggart, 2018). I have deliberately chosen to forego a higher income elsewhere in order to enjoy a more flexible schedule with fewer total hours worked in a year, and I suspect many in academia have made a similar calculation. Students are constantly reminded of the college–job–wealth life progression, but we as faculty need to remind them that *walking away* is an equally viable option as well. There are numerous examples of purposefully withdrawing from the typical, non-sustainable lifestyle (Nearing & Nearing, 1989; Klein, 2007; Kingsnorth, 2013) and our students will benefit from considering these alternative views.

2. Intentionally not owning a smartphone. This shocks most students and colleagues alike, as smartphones are ubiquitous in our culture. Though a number of benefits can be argued (countered by a lot of concerning behaviors), ultimately I worry that the numerous distractions offered by a phone would eat away at the time I have carefully protected. I recognize that going without a smartphone might be too drastic for most students and faculty. A more manageable adjustment might be to first become aware of the amount of time spent interacting with their device (and for what purpose), perhaps through simply logging their time online. Next, I would challenge them to reclaim some of their time by limiting their phone use to certain times and purposes.

3. Learning to say no to some requests (including responding to student emails during certain hours I am with my family). Like walking away from traditional consumerist choices, *saying no* is a viable alternative that can help students reclaim their time. I have built sustained blocks of thinking, learning, and writing time into my schedule, along with hours that I need to be fully present with my family. I have adequate office hours spread throughout the week to help students, and I frequently check and respond to questions via email. However, my students know, and really seem to respect, that certain times of the day I have other priorities.

4. Choosing my primary form of transportation to be by bicycle instead of driving. Yes, riding has a myriad of positive health, environmental, and social impacts, but at the core of why I ride is a desire to respect and value my own time.[3] The bicycle imposes physical limits with respect to being at different places at different times, so that I have to truly consider what opportunities I find meaningful. This sifting and winnowing of my calendar forces me to think before saying yes, and to say no more often than I otherwise would.
5. Volunteering for a number of campus and community organizations and requiring (in one course) students do the same.[4] Hearing and learning from other people's perspectives and experiences, building resiliency through our social networks, and remembering how to help others are some of the skills that volunteering fosters. These skills are crucial for environmental stewardship and fulfilling our true potential as human beings. Furthermore, volunteer efforts offer a glimpse of truly meaningful activities one can undertake when not spending all of one's time pursuing traditional consumerist ambitions.

Walking away, saying no, opting to live lower on the consumption chain— these all appear to involve giving something up. And to be very clear, I am indeed suggesting that we teach students that it is acceptable to avoid certain jobs, opportunities, and consumer goals in order to better value their time, which I argue will allow for deeper engagement in the natural world and more sustainable choices.[5] In the language of the Deep Adaptation framework, this giving up is termed *relinquishment* and on the surface it is a hard thing to do. We do not live in a culture of relinquishment, of doing without. We live in a culture that constantly tells us we can and should have anything we want, which makes the entire idea of relinquishment hard. What might emerge in our lives by *choosing* to relinquish certain things, to exhibit restraint? Besides a sense of agency and empowerment in the running of our lives, might we find more fulfillment? Certainly, we would find more time, time that we could then fill in ways of our *own* choosing. From any product life cycle analysis, the very act of relinquishing even some consumerist goals is in itself environmentally right; the newfound time one has because of this relinquishment can act as a positive feedback loop resulting in better and better environmental understanding and sustainable living.

Volunteering is different from the first four suggestions. Here, it appears that I am suggesting relinquishing some of the time students so desperately need to reclaim, but that is not quite the case. Deep Adaptation asks us to consider *what should be relinquished* and, for the reasons listed earlier, volunteering one's time should be *restored* if lacking, not relinquished.

Finally, it is not enough to mention these suggestions or similar examples from a faculty member's own life to students or advisees; a key component is clearly articulating how specific actions can lead to valuing one's time. Then, we can challenge our students to take this time and become more active protagonists in their own lives, including more actively pursuing their sustainability education through deep learning experiences (Warburton, 2003). Climate change, environmental issues, consumerism, work, fostering deeper and more meaningful relationships, personal responsibility, social justice, racism, poverty—all of these are intertwined. Thoughtfully responding through how we choose to live our lives requires time, which thus requires that we value our time. It may not be possible to teach students to value their time (some might have to learn the hard way as I did), but it is worthwhile to try.

III.   USING OURSELVES TO MODEL THE LEARNING PROCESS

Many professors are fond of paraphrasing Margaret Mead, saying that we teach "how to think, not what to think," which is an admirable goal. A major component of thinking well is learning well, and that process began long before entering the university. In college, students have undoubtedly encountered new material that they have to engage with—to learn—using previous learning techniques they brought with them and perhaps trying new ones. Students taking a first-year experience course might take a quiz to ascertain their learning style, and student-learning centers often provide additional coaching. However, typical student-only learning strategies can overlook a wonderful resource, namely their professors who (with earned doctorates or master's degrees) presumably all know how to learn new material effectively and efficiently. *I propose that we model the learning process for our students by placing ourselves in situations where we are not the subject-matter experts anymore but are instead in a student role.* Professors often observe each other in this role as learner—at conferences, campus colloquia, and in the peer-review process, for example. These are all crucial pieces of an academic livelihood, and I argue that it is important that students have access to this aspect of their faculty. Furthermore, different professors learn in oftentimes very different ways, so students need to observe a number of their professors in the role of learner in order to best inform their own development.

On most days, when I step into a classroom I have more depth, more breadth, and more experience with respect to the day's lesson than anyone else in the room—as I clearly should, it's my job as a professor. Entering the classroom and not being the most knowledgeable person about that day's topic will likely challenge many faculty members. However, by stepping out of the expert role and into the learner role, we can demonstrate our own learning styles for our students, which they can integrate into theirs as applicable. Placing faculty in a learning role might also foster a deeper connection with students since both are now participating in the learning experience from the same perspective. I try to model my own learning process for my students in a few ways.

1. Team-teaching courses with other Faculty well outside my discipline. I co-teach (we both lead and participate in *each* class meeting, which differs from many team-teaching models) *What If They're Right? Individual Responses to Climate Change* with a philosophy professor (Jenkins & Stone, 2019) and *Literature of Ecology* with an English professor. I bring my scientific background to both courses, but I have no formal training beyond general-education type courses in either of my teaching partners' subject areas (and vice versa). When my teaching partner introduces material that is unknown to me, I switch into the role of student. I ask questions, ask for clarification, ask for additional reading, try to connect the current idea to a previous example, and do whatever else is necessary to satiate my curiosity. In short, I learn in front of my students so they can see the process, a process informed by many more years of experience than they have. Moreover, the students are able to see my teaching partner and I in thoughtful dialogue with each other, so they see two, often very different, learning styles, as well as a model of civil discourse.

2. Using a primary text from outside my field. For example, the last time I taught the freshman level general-education course *Energy & Society* I used a renewable energy text that focused more on historical energy transitions from a business perspective (Usher, 2019). I provided additional scientific detail to each chapter in order to meet the science-related outcomes of the course. Equally important, though, was sharing my own engagement with the material in the text, including questions I developed while reading, resolving those questions, and additional resources I had to look up to understand certain points.

3. Inviting practitioners from the local sustainability economy or researchers into my classes. Many faculty members invite guest lecturers into their classrooms and I wholeheartedly support this effort. For example, when learning about solar power, it is hard to imagine a better classroom resource than someone who works in the solar industry and has direct installation experience, and who can speak to the economics of residential solar, environmental aspects, and policy considerations in a very personal way. To further enhance this experience, I suggest that the faculty member consciously adopt the role of learner and engage the presenter so as to model the learning process for students.

In the framework of Deep Adaptation, more capable learners build individual and community *resiliency*. Students that have broader learning skills will be able to better adapt to a world transformed by climate change. In the coming century, it is very likely humans will have to learn new skills, new jobs, new ways of living, and respond in new ways to still unknown climate change effects. Modeling the learning process for our students gives them additional strategies for their own learning and helps them develop a deeper learning skillset.

IV.   SUSTAINABILITY CONTENT CHOICE

The final personal faculty behavior I suggest incorporating in our courses is actually an encapsulation of all of our behaviors. *I propose that our course content include as many personal sustainability choices from our own lives as possible; importantly, we must include both successes and failures.* Content choice will reflect the course outcomes, but it is challenging to think of a course that cannot include a number of relevant environmental examples. As faculty already know, real-life examples always enhance presentation of coursework. And as the triple bottom line approach and the SDGs remind us, including sustainability-related examples naturally promotes interdisciplinary thinking as well. Content that I utilize from my own personal and professional life includes:

1. Analyzing energy-related data collected from my own home and from our university. Some of these data include electrical consumption after installing a hybrid hot water heater (Stone, 2018), household propane usage, residential and campus solar proposals, and campus food waste. We analyze all from multiple perspectives.
2. Participating in campus Earth Day events alongside my students as a required class assignment. We also volunteer together at a local agricultural fair that exposes students to organic agriculture, renewable energy, environmental policy and activism, and other facets of environmental stewardship.
3. Extended work in the campus garden as a required course component (Berry, 2017). I link this work to my own home garden, food preservation, industrial and local agriculture, conventional versus organic, embodied energy, and food choices. The garden supplies produce to the campus dining center and a local food bank, so we always talk about food insecurity, poverty, and the complicated links to climate change and environmental stewardship.
4. Divestment efforts. I have attempted to divest my retirement savings from investments that support fossil fuel extraction and have helped organize

for the inclusion of fossil fuel divested investment options in the faculty 403(b) plan.

5.   Previous work experience. Before entering academia, I was a submarine officer and qualified to operate a nuclear reactor. I use this experience to teach my students about nuclear power and then to provoke discussion about nuclear power as a "green" energy choice.

These examples, and others from my personal and professional life, all reflect choices that I have made. These choices are what I personally, after careful consideration, have decided to relinquish, restore, or make more resilient. These choices are my answers to the question *what can I do?* Some have been environmentally right, and some have not. My hope is that I have made my students more deeply consider their own choices, while also making their coursework with me more meaningful.

V.   CONCLUSION

In many of the courses I teach, the kernel of the semester evolves into the question, *what do you value?* Oftentimes, students value the same things that most of us do: family, friends, community, self-sufficiency *and* interdependence, education, clean water, health, good food, access to these things for all. We often spend the semester circling this question within the course context, while also drawing in multiple perspectives. Ultimately, some students recognize that what they value does not correspond to how they live—and that is when progress begins.

In answering the question, *what sustainability behaviors from my own life do I model for my students?*, I am really answering the question, *what do I value?* To the best of my ability, I value my time and believe that is the most important action I can pass onto my students.[6]

I also try to model how to learn and then include examples—both successes and failures—from my own life, but time is the critical factor. Without time, we cannot even determine the questions, much less the answers.

NOTES

1.   The phrases "climate change" and "global warming" concisely describe what is happening. However, I prefer "climate emergency" since the word "emergency" clearly articulates the magnitude of the situation while also calling us to think about what might *emerge* as we traverse the coming century. I use all three phrases interchangeably in this chapter. It should be clear that I feel the climate emergency to be the most pressing environmental issue presently, but it is certainly not the only one. Focusing on carbon dioxide, greenhouse gases, or even the climate emergency alone is too much of a simplification; responding to the climate emer-

gency involves engaging with other environmental (and social, and economic, etc.) issues *simultaneously* (Eisenstein, 2018b).

2. I feel disingenuous using the singular "I" without acknowledging the support I receive from my family, friends, colleagues, and community in all of my pursuits. This support hints at the vast interconnectedness of nature.

3. Equally important in this transportation decision is my desire to more deeply connect with the *place* I live in, namely by removing the glass windshield that so often separates us from more fully experiencing the world around us.

4. Until recently, I did not receive any compensation or course release time for these volunteer efforts (and the "credit" faculty receive for service generally falls far below that for teaching and research during the faculty review process). How to best support faculty (especially junior faculty) who want to engage in this type of sustainability work is a challenging problem.

5. Some might view the choices around relinquishment as resulting from my privileged position as a white, male professor. This is a fair question to ask, and I am still learning and grappling with how systemic inequalities inherent to our country have affected my own life and choices. A discussion of privilege deserves more space than I have here, but I will mention two points. First, valuing one's time is a concept that cuts across socio-economic backgrounds. The approach will look different, but the fundamental idea that we need to help our students value their time remains the same regardless of background. Second, many of our students (and likely the reader as well) are in a privileged position where thoughtfully relinquishing certain behaviors can lead to the time necessary to more deeply engage in questions of sustainability and environmental stewardship.

6. "Valuing time" is my (current) answer, but I hope other faculty will pursue the broader research question that asks what we should teach in order to encourage environmental stewardship. For example, Scott Russell Sanders (2020) makes a powerful argument that imagination is a crucial ingredient for environmental action, one that resonates deeply with me personally.

REFERENCES

Bendell, J. (2020) *Deep adaptation: A map for navigating climate tragedy*, July. Retrieved September 18, 2020, from http://www.lifeworth.com/deepadaptation.pdf

Berry, W. (2017) *Think little*, The Berry Center, March. Retrieved September 18, 2020, from https://berrycenter.org/2017/03/26/think-little-wendell-berry/

Coplan, K. (2019) *Live sustainably now: A low-carbon vision of the good life*. New York, NY: Columbia University Press.

Eisenstein, C. (2018a) *A new story of climate change: New frontiers* [video file], May 2. Retrieved September 18, 2020, from https://charleseisenstein.org/video/a-new-story -of-climate-change-new-frontiers-2018/

Eisenstein, C. (2018b) *Climate: A new story*. Berkeley, CA: North Atlantic Books.

Hawken, P. (2018) *Drawdown: The most comprehensive plan ever proposed to reverse global warming*. London: Penguin Books.

Jenkins, N. & Stone, T. (2019) Interdisciplinary responses to climate change in the university classroom. *Sustainability: The Journal of Record, 12*(2), 100–103. https://doi.org/10.1089/sus.2018.0033

Jensen, D. & McBay, A. (2009) *What we leave behind*. New York, NY: Seven Stories Press.

Kanashiro, P., Rands, G., & Starik, M. (2020) Walking the sustainability talk: If not us, who? If not now, when? *Journal of Management Education, 44*(6), 1–30. https://doi.org/10.1177/1052562920937423

Kingsnorth, P. (2013) Dark ecology: Searching for truth in a post-green world. *Orion Magazine, 32*(1), 18. Retrieved September 18, 2020, from https://orionmagazine.org/article/dark-ecology/

Klein, E. (2007) Win the rat race by not running it, *Los Angeles Times*, November 24. Retrieved September 18, 2020, from https://www.latimes.com/archives/la-xpm-2007-nov-24-oe-klein24-story.html

MacFarlane, R. (2019) *Underland: A deep time journey*. New York: W.W. Norton & Company.

Morton, T. (2013) *Hyperobjects: Philosophy and ecology after the end of the world*. Minneapolis, MN: University of Minnesota Press.

Nearing, H. & Nearing, S. (1989) *The good life: Helen and Scott Nearing's sixty years of self-sufficient living*. New York: Schoken Books.

Sanders, S.R. (2020) *The way of imagination: Essays*. Berkeley, CA: Counterpoint Press.

Stone, T. (2018) Five-year post-installation review of a heat pump water heater. *Spire: The Maine Journal of Conservation and Sustainability, 3*. Retrieved September 18, 2020, from https://umaine.edu/spire/2019/09/18/stone/

Taggart, A. (2018) *How work took over the world* [video file], May 8. Retrieved September 18, 2020, from https://www.youtube.com/watch?v=F7UonZl-Gis&t=62s

United Nations. (n.d.) *Sustainable development goals*. Retrieved September 18, 2020, from https://www.un.org/sustainabledevelopment/sustainable-development-goals/

Usher, B. (2019) *Renewable energy: A primer for the twenty first century*. New York: Columbia University Press.

Warburton, K. (2003) Deep learning and education for sustainability. *International Journal for Sustainability in Higher Education, 4*(1), 44–55. https://doi.org/10.1108/14676370310455332

13. The story of a sustainability cabin: Muir vs. Pinchot

Van V. Miller

INTRODUCTION

In an enlightening biography of Alexander von Humboldt, Wulf (2015) detailed how von Humboldt came 'to invent' nature through his travels, research efforts, and writings. As she interpreted him and his work, von Humboldt was the individual who linked the dots of the natural world and revealed its interconnected essence and its innate beauty (the incredible plates in his books bear witness to nature's aesthetic value). This interconnected essence still dominates ecological perceptions and thinking about life on Earth. For example, the Covid-19 pandemic has been explained in causal terms as a set of connections among loss of wildlife habitat putting stress on wild species and promoting easier transmissions of mutating viruses across species, including humans. Though von Humboldt noted the ecological destruction by humans that he observed in his travels, he did not achieve his scientific eminence due to discussions about those observations. Instead, he focused on the beauty and harmony of nature, and discussions about destruction would come later from those labeled as naturalists.

By the early 20th century, two intellectual camps of naturalists had emerged—one dedicated to preserving nature and the other committed to conserving it. The preservationists, represented by John Muir (Brinkley, 2009), appreciated greatly the beauty and grandeur of the natural world and wanted it left intact and unscathed. On the other hand, conservationists, like Gifford Pinchot, advocated for nature's utility through multiple uses (Brinkley, 2009). Beauty, being less appreciated, was forced to compete with other values perceived as more useful to humans. This controversy between aesthetic vs. utility values has become a mainstay of the ecological and sustainability causes, and its likely trajectory can be seen clearly in the Hetch Hetchy dam battle in which Muir and Pinchot staked out their respective positions and struggled vociferously for years (Worster, 2008). (Given recent revelations [Brune, 2020; Worster, 2020], neither man should be viewed as an exemplar of human values beyond those applicable to the ecological domain.) In the end, Muir lost, and the dam was built, assuring San Francisco of a long-term water supply. The preservation of the Hetch Hetchy valley's beauty, much appreciated by Muir, was far less important to most Californians than the utility of the water in the reservoir that hydrated the city yet entombed the valley.

As highlighted above, a paradigm for explaining and discussing a sustainability endeavor can be anchored by three main concepts—*beauty* as perceived in nature, *interconnectedness* as found in nature, and *utility* as taken from nature.[1] These three points will be elaborated upon herein and provide the conceptual and organizational underpinning for my discussion of a sustainability cabin project that began in 1984 and is still ongoing.

In August 1984, after successfully defending my doctoral dissertation at the University of New Mexico, I traveled 165 miles north to Chama, NM and purchased a lot on a ridge in a mountainous subdivision outside of the village. This area had become known to me several years earlier on a sightseeing trip in the early fall when the aspen leaves turned a golden yellow and contrasted vividly with the green needles of the pine and fir trees that predominate in these mountains. Its natural *beauty* overwhelmed me then and still does today. Walking among the trees in the forest offers a serenity and spiritual peace that for me cannot be duplicated outside such a setting. Latching on to this *beauty* became a prime motivator for the decision to construct a log cabin in these woods perched high above the upper Chama River valley and the continental divide farther west.

But constructing a cabin to appreciate the *beauty* of nature required that I learn about both the *interconnectedness* and the *utility* of nature and their inherent conflict with its beauty. Neither *interconnectedness* nor *utility* can be ignored because each has an impact on decisions related to sustainability that realistically require tradeoffs, generally less than ideal. However, this notion of tradeoffs as a mandatory sustainability strategy did not emerge (Mintzberg and Waters, 1985) in my thought and practice until late in the 1990s when I started researching and writing about ecological sustainability. As illustrated with Figures 13.2 and 13.3, most of the discussion in this chapter regarding *interconnectedness* and *utility* (with their tradeoffs) revolves around issues of solar energy and water. Each has its own particular story and unquestioned place among the Sustainable Development Goals (SDGs). In addition, there is another narrative about consideration of wood choices because their selection depends on trust in supplier or third-party certification that the trees were harvested in a sustainable manner. Without such trust, calculations of *utility* cannot be made. Nevertheless, the cabin being in the American Southwest provides a story of water and drought that illustrates well the 'under threat' *interconnectedness* among water, the forest, and human habitat therein. Given current climatic conditions, the naturalist debate over preservation vs. conservation may not be relevant any longer, regardless of one's ecological stance, and I will comment on that bleak outlook in the conclusion.

In the next section, I briefly explain the conceptual paradigm for organizing and presenting the cabin narrative and then I discuss the three resource factors analyzed with the paradigm. Those factors—electricity, water, and wood—provide the empirical foundation for illustrating and discussing the paradigm. Where opportune, I also note in the discussion how work on the cabin project provided me with teaching material for the classroom. It permitted me to establish legitimacy in the minds of the students. Finally, I link these factors to sustainability teaching experiences at both the undergraduate and graduate levels and conclude the chapter with brief comments about the challenges of teaching sustainability in an academic setting.

CONCEPTUAL PARADIGM

As proposed above, a concise paradigm seems more appropriate for organizing an essay about a professor's sustainability endeavors than an abstract theory with hypotheses. That said, word limitations restrict me to a mere summary justification for the paradigm. Figure 13.1 below shows the three major constructs that anchor it, and the shading reveals the proposed priority in practice (dark, lower and light, higher) of the anchors. This prioritizing seems normative yet reasonable due to its salience as a personal motivator to engage in sustainability undertakings. The task of preserving or conserving nature becomes more feasible when individuals are convinced that it deserves being sustained due to its awe-inspiring beauty. Until that occurs, decisions and actions about tradeoffs to sustain nature are mostly abstract discussions based on functional or survival imperatives.[2] The personal motivation provided by nature's beauty can be clearly seen in the definitive work of von Humboldt, Muir, and Pinchot. An appreciation for natural beauty may not be biologically innate to humans as Wilson (1984) argued, but it does hold many of us in awe. Regrettably, I often found it lacking in many students taking sustainability courses and will return to this point in the Teaching discussion. The validity of interconnectedness and utility as sustainability anchors can be accepted *prima facie* given their ubiquity in sustainability research, practice, and teaching (see an early discussion by Buchholz, 1993, regarding The Wilderness Act of 1964 and nature's relatedness and utility).

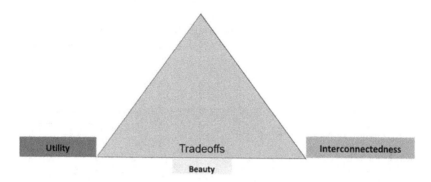

Figure 13.1   Tradeoffs among utility, beauty, and interconnectedness

Energy/Electricity

The cabin in the woods was never intended to be a short-term residence like Thoreau's shack on Walden Pond; instead, I viewed it as an abode where family and friends could gather annually and enjoy each other's company in a 'natural' setting. Knowing that neither family nor friends would delight in sleeping under the stars as the early naturalists did, the choice of a modern residence seemed obvious. A log cabin structure within a forest of pine, fir, and spruce trees interspersed with Gambel oak offered a good option for the house and appealed to my woodworking avocation.

Choosing the wood option was relatively easy compared with those required for deciding how to power all the modern devices in a house. Those decisions were somewhat constrained by the fact that the lot for the house was located off-grid. A visit to the offices of the local electric cooperative revealed that the estimated cost to extend the grid to the lot would be $50,000 (in 1988 US dollars), which was a prohibitive amount for me. Thus, photovoltaic energy became the choice for electrical power (see Figure 13.2). To deal with the high cost of photovoltaic energy, the quest became one of minimizing electrical usage. One way to accomplish that involved substituting other energy sources for photovoltaic electricity where feasible. The two most likely substitutes were propane and wood, but the utility of each required tradeoffs with beauty and interconnectedness. Although wood from trees can debatably be considered as renewable energy, it clearly diminishes the natural beauty of the forest and its burning contributes to global warming/climate change from the production of various gases, e.g., CO_2 and soot (Gardiner, 2018). Propane, a fossil fuel, is not renewable, and its combustion is also interconnected to global warming/climate change. The tradeoffs made were: use wood only for heating when the outside temperature drops well below freezing; utilize a 99% efficient propane heater in the fall and spring; cook on a propane gas stove; heat water with a tankless propane hot water heater; and rely on a highly efficient propane refrigerator–freezer. By selecting the propane-powered devices carefully, I was able to reduce gas consumption and minimize the expenditures for solar electricity.

Figure 13.2   Photograph: Solar panels with backup generator

The expenditures for the propane appliances came early in the project (1990s) and allowed me to wait for improvements in both the efficiency and the cost of a solar-electric system. In 2006 when I first installed the solar-electric system, the eight original panels represented the major expense, approximately $800 each. Today (2020), each panel costs about one-fourth that amount, and the reductions in panel costs over the years has allowed me to add more panels and rely less on wood and propane for energy. The costs of the other components have remained more or less the same or increased during the past 15 years.

Most knowledgeable observers recognize that the cost reductions for solar arrays have been driven by Chinese producers with commanding competitive positions. In terms of sustainability teaching and program development, I realized in 2010 as we launched our new undergraduate minor that these personal experiences with photovoltaic energy could be used in the classroom to establish legitimacy with students from across the campus.

Water

In the American Southwest, drought represents an ongoing fact of life. It comes; it goes; and, it returns. There are wet decades, and then there are dry decades. Failure to recognize this can lead to calamity as the historical record shows. Generally, rural New Mexico lacks water districts as one finds in more urban states. Thus, rural homeowners must be self-reliant for water. Nevertheless, in Michigan where our students were located the situation was quite different. The lower peninsula is surrounded by water, and rivers are ubiquitous. However, this fact seems to have created an indifference to water in the eyes of many students and became the decisive factor in the decision to highlight water in our course offerings. My experiences with water at the cabin gave me additional legitimacy before students in class discussions about water.

In a rural area, the three main methods for obtaining water are: pump it from an underground source, harvest/collect it from rainfall, or haul it in from an external source. There are drawbacks to each method, and an individual may rely on a combination of them. Though water usage is regulated by the State, it does allow homeowners to drill a well and pump an ample quantity for domestic consumption. However, this method carries risks (quantity and quality of water), costs thousands of dollars if the water table is far below ground, and depends ultimately on the underground water source continuing to exist. In a drought-prone area, natural recharge of the source becomes a critical issue and requires expensive analyses to gauge. In light of these issues and local conditions, I chose to haul potable water from a nearby state fish hatchery, which had traditionally provided local residents a free-water tap to be drawn from for household consumption.

In 2018, the free-water era ended. The tap at the hatchery was shut off. The aquifer that provided water to the spring the hatchery depended on was slowing down. The ranchers/growers who actually owned the water rights for the water had an insufficient quantity for their agricultural needs and ordered the hatchery to stop supplying water to the tap and its local users. The order was drastic and compelled several hundred of us to consider our alternatives. I opted to haul, when necessary, purchased water from a much farther distance and to expand the rainwater collection system that had been started several years earlier (see Figure 13.3). With the end of free water, the

Figure 13.3   Photograph: Water collection and storage expansion (under construction)

utility taken from nature became a necessary factor to consider and resulted in my decision to collect rainwater with backup from long-distance hauling if needed. Nevertheless, this water decision for human consumption must be interconnected with water in general, i.e., a greater shortage of water due to climate change and its impact on the wooded forests plus its long-term effect on the aquifer. Water, essential for life on Earth, cannot be treated as just another resource to be sold and bought in the marketplace. Although markets have a place in water transactions, those making such transactions should recognize that subjecting Earth's sustainability to markets can jeopardize life itself (Miller and Pisani, 2018).

Wood/Sustainability Certification

As already noted, tradeoffs and decisions about utility for sustainability endeavors often require third-party certification. Were the ocean fish harvested in a sustainable manner? And were the eggs collected from range-free chickens? These are the types of questions certification addresses. The same can be asked about wood. Were the trees logged in a way that sustains the forest? The provider of the spruce logs for the cabin told me they had been harvested from trees 'dead and dried on the stump'; thus, their logging should not have contributed to deforestation. But, do I know the accuracy or truthfulness of his statement? I do not. Third-party certification can remedy this problem if there are agreed-upon standards and trustworthy certifiers. The value and efficiency of third-party certification I can attest to by describing the production trail for the ash ceiling boards used in the cabin. This laborious example was also presented in the classroom to augment my legitimacy in a critical area impacting sustainability decision making.

To clear a building site for a home in an upper Midwestern state, I contracted to have a half-dozen ash trees cut down and sawed into lengths of 8–10 feet. After supervising this operation, I then transported them to a sawmill where they were cut into rough boards. Next, I took them to a kiln where they were dried before storage and eventual transport to the cabin site in New Mexico. Once there, they were cut, planed, shaped, and sanded by me for placement in the ceiling. Without a doubt, I can determine the sustainability impact these boards had upon the natural environment. Though the effort provided novel experiential learning, I do not wish to repeat it. We cannot expect a general contractor or a DIY person in Home Depot buying some boards to engage in such certification activity. Hence, we must rely on third-party certification of the businesses and products we use (e.g., the GRI, Global Reporting Initiative, 2021, for firms in general, and the FSC, Forest Stewardship Council, 2021, for those who produce/sell wood or wood-derived products) and direct our efforts to improving their reliability and educating others about the certification process.

In the description and discussion for each of the resource factors, I have noted in particular how each was used to augment my sustainability legitimacy in the minds of the students. These personal experiences, along with numerous class assignments and trips (explained below), represented the substance of our courses. Did the cabin project enhance my legitimacy? I cannot be certain, but I did observe that students who had intense interest in sustainability would later call or e-mail me for references or information.

TEACHING AND CONCLUSION

Beginning in 2008 and for the next five years, I was the Program Director on two grants from the US Department of Education. Both grants were for the purpose of creating and establishing innovative programs to integrate the international business curriculum with the concepts and tools of sustainable development. Involvement in both the creation and implementation of these programs gave me first-hand experience in teaching the new curriculum to undergraduate and graduate students. Upon reflection, I remain convinced that the cabin sustainability project gave me greater creditability and legitimacy in the minds of my students. It gave me first-hand experience with the SDGs, particularly Goals 6 and 7—clean water and clean energy. The SDGs were also covered in our readings and class assignments, but the targets associated with each goal were only emphasized when a student had a project that made them relevant. In line with these, I would usually mention the cabin project early on in a course and then refer to it where apropos throughout the semester. In addition, I developed a short cabin presentation for recruiting new students to the programs. In my mind, the project demonstrated that the professor had lived what he researched and taught regarding sustainability.

In terms of the three resources discussed above, we would primarily rely on outside experts to increase our understanding of particular sustainability issues, for example, battery technology or water chemistry. In addition, we would make site visits to facilities that engaged in wood manufacturing, water bottling, and the generation of electricity from renewable energy. These experts and visits served as another source of legitimate knowledge for a particular sustainability issue that went beyond what we learned in the classroom. Their shortcoming was they failed to develop the interconnectedness concept of nature. Each expert or visit only reinforced a particular issue. Nevertheless, they prompted good discussions whenever the students in the group were sufficiently motivated to question seriously the expert or practitioner talking with us.

This last point brings me to what I noted in the Conceptual Paradigm section and asks a question that I still find unanswerable: how does a professor instill in the students an appreciation of nature's awesome beauty? The answer becomes even more complicated when the class is online, which is where our graduate teaching of sustainability occurred. This format precluded the use of plant visits and guest professors from across the campus. To make it more realistic, we incorporated a sustainability business simulation, an early version of Realia's Green Business Lab (2021), into the course. It worked very well with graduate business students, given their maturity, the lab's strong emphasis on utility and the use of different components in manufacturing.

Our undergraduates, however, did not have exposure to the lab and lacked a mindset for beauty. The required liberal arts courses did not address it, and my business-oriented classes did not deal with it. In hindsight, I would revise the program and ask the philosophy and fine arts faculty if a course on nature's beauty could be developed. Although such an approach perpetuates the discipline-oriented silos of the university, it does recognize the problem and attempts to remedy the issue in spite of a reward system that discourages integrative endeavors across disciplines and colleges.

In conclusion, projects that focus on sustainability can be very beneficial in a professor's own experiences and professional labors. In my personal life, the water challenge discussed above has made me very conscious of water usage, and I limit its use whenever possible. However, when away from the cabin I have noticed that my concern about water usage tends to diminish. Correcting that shortcoming requires thought and discipline. Professionally, the cabin project has allowed me as an instructor to bring real-world problems into the classroom and illustrate important issues in ecological sustainability. However, class projects usually assume that the problems can be solved if we rearrange the pieces of the manufacturing process or supply chain differently. In 2020 with its mega-wildfires and pandemic across Earth, such an assumption looks increasingly naïve. When I consider the issue of wildland fires and their interconnections with water, forests, and global warming/climate change, tweaking the pieces, e.g., better forest management, seems like rearranging the deck chairs on the *Titanic*. The problem is much greater and more complicated than that and appears to mandate changes that most of us will be loath to adopt. Perhaps, the complicated changes required for sustainability could be handled better if we struggled more in our analyses and decision making by reaching for the stars of Muir's 'preservation' while minimizing the necessity of Pinchot's 'conservation'.

ACKNOWLEDGMENTS

Thanks to my wife, Patricia, whose labor and support enabled me to undertake this project and who has never tired of the solar clothes dryer, aka clothesline. A debt of gratitude is owed to Tom and Chuck for their many visits and efforts, and to the late Joel, who readily understood utility and the need for tradeoffs.

NOTES

1. To define these three terms, I consider each showing a human value that demonstrates personal and social preferences influencing decisions and actions. Their weights (non-additive) in decision making do and will vary across time and circumstances. Beauty denotes an ascribed aesthetic value that represents both

the awesomeness (lightning storm) and grandeur (Grand Canyon) of the natural world. Interconnectedness reveals the value of necessary evolutionary links between/among natural phenomena (climate, rain/snow, and forest vegetation) that is primarily beyond human influence, but not always (see Egan [2006] for an account of how humans altered the climate and topography of the US High Plains in the late 19th and early 20th centuries). Utility can be described by the discrete values, often monetary, attached to natural resources that when taken from their innate setting provide usefulness to humans. Of the three, interconnectedness is the one that relates best to the sustainability concept.

2. One would think that functional imperatives, esp. those dealing with survival, would be sufficient to motivate individuals to engage energetically in sustainability efforts, but often that does not seem sufficient. The logical case for such efforts presumes that others view the situation as sustainability advocates do and that those others believe their actions will have the desired effect, assuming they agree that the effect is desirable. After witnessing the tragic US events of 2020, I am not convinced that the issue of survivability assures the expected outcome. Perhaps, as suggested herein through the examples of Muir and Pinchot, an increased appreciation of nature's beauty will accomplish more than dire warnings about earth's survivability.

REFERENCES

Brinkley, D. (2009). *The wilderness warrior: Theodore Roosevelt and the crusade for America*. New York, NY: Harper Perennial.

Brune, M. (2020). Pulling down our monuments. *Sierra Club*, 22 July, accessed at https://www.sierraclub.org/michael-brune/2020/07/john-muir-early-history-sierra -club on 17 January 2021.

Buchholz, R.A. (1993). *Principles of environmental management: The greening of business*. Hoboken, NJ: Prentice Hall.

Egan, T. (2006). *The worst hard time*. Boston, MA: Mariner Books.

FSC. (2021). Forest Stewardship Council, accessed at https://us.fsc.org/en-us on 9 June 2021.

Gardiner, B. (2018). Is your wood stove choking you? How indoor fires are suffocating cities. *The Guardian*, 22 February, accessed at https://www.theguardian.com/cities/ 2018/feb/22/wood-diesel-indoor-stoves-cities-pollution on 28 May 2021.

GRI. (2021). Global Reporting Initiative, accessed at https://www.globalreporting.org on 9 June 2021.

Miller, V.V. and Pisani, M.J. (2018). Sustainability science and water usage: Science as a method for the corporate governance of natural resources. In G. George and S.J.D. Schillebeeckx (eds), *Managing natural resources: Organizational strategy, behaviour and dynamics*. Cheltenham, UK and Northampton, MA, USA: Edward Elgar Publishing, pp. 245–270.

Mintzberg, H. and Waters, J.A. (1985). Of strategies, deliberate and emergent. *Strategic Management Journal*, 6(3), 257–272.

Realia. (2021). The Green Business Lab, accessed at https://www.gblsim.com on 11 October 2020.

Wilson, E.O. (1984). *Biophilia*. Cambridge, MA: Harvard University Press.

Worster, D. (2008). *A passion for nature: The life of John Muir*. Oxford: Oxford University Press.

Worster, D. (2020). John Muir biographer: He was no white supremacist. *California Sun*, 30 July, accessed at https://www.californiasun.co/stories/john-muir-biographer -he-was-no-white-supremacist on 17 January 2021.

Wulf, A. (2015). *The invention of nature: Alexander von Humboldt's new world.* New York, NY: Alfred A. Knopf.

PART III

Curriculum development in sustainability education

14. An ecocentric radically reflexive approach to walking the "Earth System talk" in sustainability education

Melissa Edwards and Wendy Stubbs

INTRODUCTION

As two sustainability-focussed academics, one of our main fields of influence occurs in our practice and process of curricula development. Curricula provide a framework for seeing, understanding, evaluating, analysing, and innovating. They are imbued with ontological and epistemological assumptions that represent the dominant values and ways of doing and being within a discipline or field of study. For instance, Sterling (2001) has posited a distinction in two dominant worldviews underpinning education. A conventional curriculum is based on a reductionist ontology and an objectivist epistemology where knowledge is generated in compartmentalised disciplines. Learning excellence is measured by the relative success in the transfer of knowledge from teacher to learner, whereas an emergent learning for sustainability approach is based on an integrative ontology and a participative or critical subjectivist epistemology. Learning is "holistic, systemic, and ecological or relational" and is deeply experiential and draws on and across disciplines (Sterling, 2014). Educators for sustainability can be reflective practitioners trying to create and hold space for radical imaginations to consider future generations and the Earth System[1] that supports our livelihoods. Education for sustainability is a holistic approach that draws on and weaves across traditional disciplines (Shrivastava, 2010) to address complex and seemingly intractable issues, such as climate change, persistent social inequalities, and the regeneration of the Earth System – issues encompassed in the Sustainable Development Goals (SDGs) (United Nations, n.d.). As educators working within this space, we write this chapter as:

1. an ecocentric radically reflective text highlighting instances in our practice where we engage in processes to enable paradigmatic change for sustainability in business curricula; and,
2. an appreciation of the processes that have worked and failed as we embarked on embedding sustainability in higher education curricula.

We describe some of the artefacts of our practice (see Table 14.1) and show how we seek to give voice and space to an ecocentric perspective, so that symbolism and new social imaginaries may permeate and enable emergence of new ways of seeing business and being businesspeople. Quite simply, social imaginaries are the way in which people imagine their existence. Rather than being fixed, we follow the view that imaginaries are polysemous; that is they have multiple meanings in different fields, and are being shaped and reshaped as people individually and collectively attach meaning to their existence (Gilleard, 2018). Therein, radical human imagination can posit, experiment with, and determine and create new ways of attaching meaning to human activities as they operate within an industry, economy, and society. Imagining an ecocentric perspective can invoke different ways of valuing and instituting forms and mediums of exchange, production, and consumption, as if humans are interdependent on and intertwined with ecologies. Social imaginaries are spaces where radical imagination exercises ecocentrism through a social process, whereby individuals and collectives call into question, reflect on, and create different ways of doing business within the Earth System. Essentially this means we adopt a critical relist perspective in our classrooms. We encourage reflection on and critique of the unsustainable practices of the dominant ways of doing business and encourage students to imagine alternatives. We share our personal sustainability actions – "individual efforts that demonstrate care for self and others, and especially the planet and its biosphere, while acknowledging the interconnectedness of all living entities at all levels" – and how we enlist our students to engage in sustainability actions that create positive, collective environmental and social change (Kanashiro et al., 2020).

Through storytelling, we illuminate how we create space for Earth System talk by walking a radically reflexive path with our colleagues in the practice of embedding sustainability in the curricula at two Australian universities. Our stories reveal how our practices intertwine with shaping the social material context as well as the material artefacts and boundary objects that signify and contain the "outcomes" of curriculum development. Through a radically reflexive approach (Allen et al., 2019), we compare across our stories. We highlight our collaborations with colleagues, the artefacts of those collaborations, and the learning space we generate for students through: (1) disrupting our conventional ways of thinking about how we relate to the environment; and (2) sensitising us to our, and our students', embeddedness within the Earth System, and our responsibility to act.

Background

Radical reflexivity places humans as participants in and active constructors of our world. It promotes an ecocentric[2] approach to sustainability that "requires us to pay attention to the interrelated nature of values, actions and our social and material world. It does so by engaging students in questioning assumptions about our place in the world, the multiple and competing interests we may encounter and how we can act in responsible and ethical ways" (Allen et al., 2019, p.792). It addresses one of the main challenges in sustainability – engaging managers and students to recognise the impacts of business practices and managers' and students' own values and practices on the environment and society. This realisation needs to be connected to the need for action (Allen et al., 2019, p.786).

Radical reflexivity is about questioning our assumptions and values – what we take for granted – and the effects of business practices on the Earth System. In particular, it emphasises our responsibilities as teachers for "shaping social and organisational realities and creating responsive and responsible organisations" and students (Allen et al., 2019, p.786). Businesses are not seen as separate from the environment; rather they are embedded within it. Likewise, this approach can help both educators and students appreciate their social and physical embeddedness in the Earth System. Radical reflexivity helps teachers and students to evaluate and transform their approach to sustainability and rethink their responsibility for acting on sustainability issues.

OUR STORIES

Wendy at Monash University, School of Social Sciences

I draw upon my experiences in designing sustainability curriculum for Monash University's Master of Business Administration (MBA), Master of Business (MoB), and the Master of Environmental Management & Sustainability (corporate sustainability specialisation) to discuss how I walk the Earth System talk. My students range from believers, passionate champions, converts, sceptics, and outright rejecters. This provides both opportunities and challenges to encourage and enable students to promote positive environmental and social behaviours in their own lives, in their workplaces and become sustainability change agents (Kanashiro et al., 2020).

As students learn more from direct observations of desired behaviour than one-way information transmission (Kanashiro et al., 2020), curricula are designed to frame business and personal activity within the Earth System, and planetary boundaries (Rockström et al., 2009), and build skills and competencies to prepare students to support solutions to the complex issues underlying the SDGs.

A sustainability self-assessment using an online footprint tool is an eye-opening exercise for many students, particularly MBA and MoB students who have little understanding of their personal impact on the Earth System. It also provides a roadmap for what actions students can take to address their impacts (Kanashiro et al., 2020). I ask students to calculate their footprint using the Ecological Footprint tool (https://www.footprintcalculator.org/), and share my own footprint with the students. Students come to class prepared to speak about the components of their own footprints. Practising salient sustainability (Kanashiro et al., 2020), I explain the breakdown of my own footprint (and why, if everyone lived like me, we would need 3.4 planets) and actions to try to reduce my footprint. In small groups, students share their own footprints and discuss ways to reduce their respective footprints, before sharing their group's ideas in a larger class discussion. Other activities include guiding students through personal energy and/or waste audits.

Teachers' credibility is not only reflected in their personal actions but in the values expressed in their research and curriculum (Kanashiro et al., 2020). I utilise a sustainability worldview tool, developed in a past research project, to help shift students' mindsets (Stubbs and Cocklin, 2008) towards more ecocentric values and practices. Guest speakers reinforce how business can align with ecocentric values and implement ecocentric practices, such as biomimicry, closed loop systems (circular economy), social enterprise models, and impact investing. After being inspired by the film *2040*,[3] I invited students to watch a free screening, and discussion, of the film organised by one of these guest speakers.

Radical reflexivity is associated with questioning our individual assumptions, values, decisions, and actions, and those of organisations (Allen et al., 2019). The sustainability worldview tool helps students to challenge assumptions underpinning "business as usual" and to engage in debate around the "sustainability paradox", in which the dominant technocentric approach to business generates externalities that degrade the ecological systems and social relationships that underpin business growth. The tool helps to foster in students a deeper appreciation for the ecocentric worldview and the implications for values, ethics, attitudes, behaviours, and lifestyles (Allen et al., 2019).

Radical reflexivity is reinforced in student assignments that "require students to attempt to create change in their own sustainability behaviors, and in those of their organizations and communities" (Kanashiro et al., 2020, p.824).

Reflexive writing can help students confront previously unknown knowledge and explore multiple perspectives to help them make sense of their "entanglement in sustainability and responsibility for action" (Allen et al., 2019, p.791). In a reflective essay, my students are asked to examine their own values and worldview, the worldview and values of the industry or organisation in which they aspire to pursue a career, and the level of alignment and implications of their analysis. The final part of the student essay is a strategy and action plan for addressing sustainability issues in the chosen industry/organisation. The personal reflection essay has led some business students to change careers and pursue sustainability careers.

In another course, my students are asked to maintain a journal and record their reflections throughout the semester in response to the following questions: (1) What do you think are the key skills and knowledge required for a graduate from your study programme? What is your current understanding/ perspective on sustainability? (2) Looking back at the semester, how does your experience of the course compare to your initial expectation when you completed the Week 1 submission? Has your perspective on sustainability changed and/or been challenged in this course? What skills and knowledge have you gained and applied during this course and how do they relate to your everyday life, study programme, and chosen/future profession?

Taking a radical reflexive approach – examining multiple positions and truth claims and their consequences – has its challenges, as it is the antithesis of business schools' preoccupation with "objectivity, value neutral positioning, the separation of researcher/researched, and abstractions that detach the individual from the context/environment in which they operate" (Allen et al., 2019, p.787). Furthermore, the mainstream academic view is that, as teachers, we should engage in value-free education and not be "one-sided, biased, and dogmatic" in our perspectives on sustainability (Kanashiro et al., 2020, p.836). I walk a fine line with disaffected MBA and MoB students, who may shutdown when challenging the profit-maximisation mantra of the technocentric view and promoting ecocentric values/worldviews. It is important to be respectful and not critical of students' perspectives but provide them with the space to explore their own values and assumptions in light of the sustainability challenges and SDGs, raised throughout the courses.

Melissa at The University of Technology Sydney, Business School

I reflect on my practice over a decade of taking a learning for sustainability approach within the University of Technology Sydney, Business School, where I have collaborated to embed sustainability in the business school curricula, develop and participate in communities of practice, and create sustainability courses within the postgraduate MBA and Master's programmes.

As an academic working within the context of a business school, I take radically reflexive acts by being responsible for my own acts as an educator embedded within ecological systems and communities of practice (Cunliffe et al., 2020). I meld my personal and faculty sustainability acts through my lived experience as a business school academic by adopting a holistic "learning for sustainability" approach. I live sustainability in and across the multiple contexts of my life, as I consistently try to bring my "whole self". In a business school this means reflecting on commonly accepted ways of being and doing, where the dominant learning paradigm separates us as *economic actors* from us as *ecologically embedded humans*. The commonly accepted approach is to learn about sustainability as an add-on or case study example within a disciplinary based subject, such as accounting, marketing, or management.

Personally, I came to sustainability by connecting across various disciplines and engaging with viewpoints that challenged my worldview. I carry this insight into my practice as I consider learning for sustainability comes through inter-, multi-, or trans-disciplinarity, approaches that occur through processes of dialogue, debate, and recognising ontological and epistemological differences, even if the outcomes are seemingly incommensurate and partial. Radically reflexive acts challenge divisions between systems and contexts, giving rise to an ecocentric ontology where people realise themselves as composed of and reliant upon natural and societal systems (Starik and Kanashiro, 2013). I challenge conventional ways of doing business education in a "silo", separated from sociology and science, while also engaging with the people I encounter, my colleagues and students, to learn for sustainability in the multiple contexts of their lives as students, business professionals, and humans.

I start from a foundation of sustainability as the overarching concept within which all disciplines emerge. I ensure this is reflected in my teaching by exposing students to scientific and sociological concepts, such as the Anthropocene, Earth System science, and distributive justice, that conflict with dominant business models. For example, my postgraduate course commences with a nested systems perspective and an introduction to systems thinking as it relates to business strategy. I don't insist students adopt an ecocentric business perspective but ask them to consider how putting ecology first, would make a difference in the way they define their business mission and key activities. Challenging students to consider if and should their companies adopt global frameworks such as the Sustainable Development Goals or the Future Fit benchmarks often stimulates heated discussion. This is especially so as applying the Future Fit benchmarks framework often reveals how the impacts of their business activities falls short of their desired sustainability aspirations and intentions to operate within the Earth's Carrying Capacity. As the course progresses, I use the phase model approach (Benn et al., 2018) as a heuristic device to demonstrate the variety of ways businesses engage with a sustainability agenda and ask them to analyse any current sustainability-oriented practices in their companies. Finally, the course focusses on sustainability-oriented innovation and sustainable business models, and students are challenged to propose a new business model, strategic intervention, or business development opportunity. Students are not required to suggest radical ideas, but I take them through a series of workshop activities encouraging them to move towards the "ideal type" of a sustaining organisation, Sustainability 3.0 (Dyllick and Muff, 2016) or the Circular Economy.

My personal sustainability acts are radically reflective as I challenge myself and others to see that the system is deeply personal, and the individual is deeply part of the system. Shrivastava (2010) argues that a holistic approach to education for sustainability (EfS) requires more than a cognitive shift or changes in behaviour patterns. I take his core premise that other foundational factors such as physical and emotional engagement with sustainability issues that enable "emotional engagement and passionate commitment" are necessary. Mostly I seek to make this emotional and physical connection to the environment, in the classroom through small acts and obvious cues I use to make the natural environment visible. Through these examples, I try to establish a connection beyond cognition to deeper levels of awareness, such as an emotional or value driven connection.

An example of a small visual cue is that I choose to be a bike commuter to reduce my environmental footprint and so I often enter meetings or the classroom with a bike helmet obviously strapped to my bag. Or another small cue is the way I deliberately use images of the natural environment in my lecture slides. In teaching, I commence the class by asking students to silently reflect on a moment when they spent time in and appreciated the natural environment. I ask them to describe the natural environment and to tell me if they see anything in the room that resembles nature. Following on from this, I highlight how everything in the classroom is produced by and embodies the natural environment drawing on materials and energy. Next, I ask students to tell me where the environment is visible in business models and how it is valued. In every session, I usually find someone with a single use drink container on their desk and use this as a conversation starter to make people aware that the usual business logic does not put a proper value on nature.

I create programmes where guest speakers with varied professional backgrounds discuss the tensions and challenges they encounter when "doing" sustainability in business. And I create mixed classrooms where professionals, students, and academics collaborate on intractable or wicked problems. For example, in an undergraduate summer school, students from all disciplines come together to analyse and generate solutions for a wicked problem focussed on textile and material waste.[4] I coordinate a team of experts and practitioners from various disciplines to share information and provide feedback and coaching. We bring tangible evidence of the material waste to the classroom and students see they are part of the problem. Students do a "wardrobe audit" prior to their second class and are encouraged to bring items of clothing they have worn infrequently or only once or twice to create a pile of waste in the middle of the classroom. We ask students to consider the environmental damage in that pile, and how they might come up with different approaches.

I take a community of practice, collaborative, and networked approach to learning through collaborations with colleagues, students, and business professionals in and outside the university. For example, with colleagues we created and facilitated a national sustainability teaching and learning community of practice through creation of an open access online repository.[5] The premise is to allow experienced sustainability educators to share their best practice freely and through open access. We created regular dialogues on topics related to sustainability learning through virtual meetings allowing educators to learn from one another. Similarly, I have facilitated a faculty-wide interdisciplinary sustainability working party for the purposes of generating sustainability learning objectives that could be embedded across different disciplinary programmes. We shared a view that curriculum is living[6] and a process of dialogical co-creation rather than a stagnant artefact (Edwards et al., 2020). Over time, our practice led to various learning artefacts being embedded across and within disciplines that had been resistant to sustainability learning.

Every personal and faculty act I undertake is informed, inspired, and energised from being connected with strategies, policies, global and local research, and education networks. For example, I connected the sustainability CoP with the Green Gown Awards, the Australasian Campuses Toward Sustainability, and the Australian Association for Environmental Education. The courses I create are informed by research generated through engagement with global networks of scholars, sustainability professionals, and advocates. As I work on research related to the Earths Systems Governance I engage with colleagues, including my co-author on this chapter and other collaborators in the UK whom I met through being part of the Organisations and the Natural Environment division at the Academy of Management. Through a higher education network facilitated by the Ellen MacArthur Foundation, I have contributed to and had access to knowledge and information regarding the Circular Economy. I understand responsibility and my personal adoption of radical reflexivity through my connection into the Principles for Responsible Management Education and learning about the sustainability mindset as an inaugural member of the LEAP! group.[7] Finally, my overarching practice is strengthened and revitalised through my interactions with a community of Sustainability in Management Education (SiME) scholars (Arevalo and Mitchell, 2017).

COMMON THREADS: SMALL ACTS OF RADICAL REFLEXIVITY

Through the narratives of our practice we have told stories about how we meld our personal and faculty sustainability acts through the approaches we take as sustainability educators in two different university settings. Our stories hold common threads as we both seek transformative learning approaches to disrupt conventional ways of thinking about how we relate to the environment and sensitise ourselves to our embeddedness and shared responsibility to act (Allen et al., 2019). Table 14.1 (below) provides an overview of our personal faculty sustainability acts and those that we embed in our teaching and learning practice to encourage our students to take similar acts. Even now as we collaborate on this text, we exchange ideas and share stories about the victories, challenges, and frustrations of doing sustainability in a university context. We learn from one another and our extended networks and set ambitions and plans for next steps. For example, our recent collaboration led one author to embed the Future Fit framework in her postgraduate course. As a result, she realised the personal actions she will take to make sustainability improvements to her household such as installing solar panels. She has formed a collaboration with colleagues at her university to develop enterprise learning for climate change as well as work with senior executives on a sustainable finance plan, including divestment and decarbonisation strategy.

Common to both our approaches are the artefacts we use in the classroom to make the environment visible, to allow learners to see and make a personal connection with ecological systems and Earth Systems. We both draw on the sciences, framing our approach in systems thinking, and use simple tools such as the ecological footprint calculator to make tangible the quite abstract concepts of overconsumption and breaching of Earth's limits. In our classrooms, we take small acts of sharing our own personal sustainability stories and the acts we take outside of the classroom. We do not require our students to adopt an ecocentric approach, but we invite them to consider what it might mean for business, and we demonstrate our own sensitivities in that, as academics, we do not provide "the answers". This allows people to see that there is no magic solution for sustainability, but that we each have a responsibility to act through our professions and our personal lives.

Our stories reveal that it is not only the artefacts (e.g., the learning materials, papers, curricula, presentations, etc.) that allow people to learn for sustainability, but also the processes we use to disrupt and sensitise ourselves, our colleagues, and our students to make the environment and socio-economic conditions visible that is important. As our stories reveal our engagement in communities of practices and collaborations with colleagues from other

Table 14.1   *Summary of actions related to student activities and the*
             SDGs

SDGs	Monash – Faculty Personal Actions (Wendy)	UTS – Faculty Personal Actions (Melissa)	Student activities
7 – Affordable and Clean Energy and 13 – Climate Action	Solar panels and solar heat pump (hot water)	Green energy provider	P, C: ecological footprint tool
6 – Clean Water and Sanitation	Water tank plumbed into toilets		P, C: waste audit; energy audit
13 – Climate Action and 11 – Sustainable Cities and Communities	Drive a hybrid car and offset air travel	Bike commute and take public transport. Reduce air travel and encourage video conferencing	
12 – Responsible Consumption and Production	Vegetarian diet and grow own vegetables (give away/swap excess with friends and work colleagues)	Meat minimised diet, urban veggie growing and buy local where possible	
12 – Responsible Consumption and Production	Recycle everything	Recycle everything, choose goods with minimal or no plastic packaging	P, C. A: students examine Circular Economy business models and discuss how they can identify and support circular businesses
1 – No Poverty, 2 – Zero Hunger, 5 – Gender Equality and 16 – Peace, Justice and Strong Institutions	Donate to green organisations (e.g., Australian Conservation Foundation) and poverty reduction initiatives; invest directly in socially responsible companies adopting new business models (e.g., B Corps); invest superannuation in sustainable funds	Invest superannuation in sustainable funds	P, C: Students discuss the characteristics of sustainable businesses and what they (can) do to support sustainable businesses (e.g., student volunteering; buying certified sustainable products)

16 – Peace, Justice and Strong Institutions	Embedding sustainability worldview and business models in curriculum	Embedding sustainability worldview and business models in curriculum	G: Speakers from companies that reflect ecocentric values
			C: Analyse and discuss companies' values and worldviews across different industries
			C: Student debates: purpose of business; can economic growth go on forever?
			A: Reflective essay; journaling

Note: A = assessment; C = classroom activity; P = pre-class activity; G = guest speaker.

disciplines and professions, we are both deeply embedded in forms of boundary work. Boundary work refers to the inter-occupational interactions and exchanges as processual and relational (Ungureanu and Bertolotti, 2018). Boundary work has been studied in the context of exchanges between occupations, such as that between academics and/or practitioners, or between practitioners and/or academics with differing disciplinary backgrounds. Strategies for breaching boundaries through techniques that provoke insecurity and challenge taken-for-granted roles through experiential techniques that allow critical reflexivity are generative of new ways to understand work contexts (Ungureanu and Bertolotti, 2018).

Radical reflexivity is a critical process from where learners can view the dominant anthropocentric worldview and imagine their embeddedness within and interdependence on the Earth System that supports our livelihoods. As reflexive observers and learners, we can step outside of and become aware of dominant worldviews and our actions that uphold them. Such a position can be a catalyst for new imaginaries wherein we imagine ecocentric ways of doing business and being businesspeople and small radical acts in which ecology is valued with economy and society. Our stories are not heroic, but rather small acts that seek to make the environment visible so we and our fellow learners can continue to imagine, discover, and create ecologically centred ways of living and doing business.

NOTES

1. The relationship of humans with Earth's environment has changed significantly since the onset of the Industrial Revolution. Half of Earth's land surface has been domesticated for direct human use and most of the world's fisheries are fully or overexploited. The composition of the atmosphere is now significantly different than it was a century ago. Earth is now in the midst of its sixth great extinction event. The evidence that human activity is affecting the basic functioning of the Earth System, particularly the climate, grows stronger every year. Humans now

have the capacity to alter the Earth System in ways that threaten the very processes and components upon which the human species depends (Steffen et al., 2004).

2. Ecocentrism "promotes the inherent worth of nature rather than its instrumental value – ecosystems are viewed as having inherent worth independent from human value judgments. The ecocentric perspective draws on concepts of: egalitarianism; decentralized social, economic and political systems; bioregionalism (regions governed by nature not legislature); communalism; collectivism; and cooperation" (Stubbs and Cocklin, 2008, p.208) .

3. *2040* is a documentary that was motivated by director Damon Gameau's concerns about the planet his 4-year-old daughter would inherit. Damon embarked on a global journey to meet innovators and changemakers in the areas of economics, technology, civil society, agriculture, education, and sustainability. Drawing on their expertise, he sought to identify the best solutions, available to us now, that would help improve the health of our planet and the societies that operate within it (Gameau, 2019).

4. See https://lx.uts.edu.au/blog/2017/04/27/creativity-complexity-combines -fashion-transdisciplinary-learning/ (accessed 28 May 2021).

5. At www.sustainability.edu.au (accessed 28 May 2021).

6. At https://lx.uts.edu.au/blog/2018/04/04/sustainability-living-curriculum/ (accessed 28 May 2021).

7. At https://www.unprme.org/prme-working-group-on-sustainability-mindset (accessed 28 May 2021).

REFERENCES

Allen, S., Cunliffe, A.L. and Easterby-Smith, M. 2019. Understanding sustainability through the lens of ecocentric radical-reflexivity: Implications for management education. *Journal of Business Ethics*, *154*(3), 781–795.

Arevalo, J.A. and Mitchell, S.F., eds. 2017. *Handbook of sustainability in management education: In search of a multidisciplinary, innovative and integrated approach.* Cheltenham, UK and Northampton, MA, USA: Edward Elgar Publishing.

Benn, S., Edwards, M. and Williams, T. 2018. *Organizational change for corporate sustainability*, 4th edition. Abingdon: Routledge.

Cunliffe, A.L., Aguiar, A.C., Góes, V. and Carreira, F. 2020. Radical-reflexivity and transdisciplinarity as paths to developing responsible management education. In Moosmayer, D.C., Laasch, O., Parkes, C. and Brown, K.G., eds, *The SAGE handbook of responsible management learning and education*. London: Sage, pp.298–312.

Dyllick, T. and Muff, K. 2016. Clarifying the meaning of sustainable business: Introducing a typology from business-as-usual to true business sustainability. *Organization & Environment*, *29*(2), 156–174.

Edwards, M., Brown, P., Benn, S., Bajada, C., Perey, R., Cotton, D., Jarvis, W., Menzies, G., McGregor, I. and Waite, K. 2020. Developing sustainability learning in business school curricula – productive boundary objects and participatory processes. *Environmental Education Research*, *26*(2), 253–274.

Gameau, D. 2019. 2040. Retrieved 3 August 2020 from https://whatsyour2040.com/#

Gilleard, C. 2018. From collective representations to social imaginaries: How society represents itself to itself. *European Journal of Cultural and Political Sociology*, *5*(3), 320–340.

Kanashiro, P., Rands, G. and Starik, M. 2020. Walking the sustainability talk: If not us, who? If not now, when? *Journal of Management Education, 44*(6), 822–851. https:// doi.org/10.177/1052562920937423

Rockström, J., Steffen, W.L., Noone, K., Persson, Å., Chapin III, F.S., Lambin, E., Lenton, T.M., Scheffer, M., Folke, C. and Schellnhuber, H.J. 2009. Planetary boundaries: Exploring the safe operating space for humanity. *Ecology and Society, 14*(2), 1–34.

Shrivastava, Paul. 2010. Pedagogy of passion for sustainability. *Academy of Management Learning & Education, 9*(3), 443–455. https://doi.org/10.5465/amle .9.3.zqr443

Starik, M. and Kanashiro, P. (2013). Toward a theory of sustainability management: Uncovering and integrating the nearly obvious. *Organization & Environment, 26*(1), 7–30.

Steffen, W., Sanderson, R.A., Tyson, P.D., Jäger, J., Matson, P.A., Moore III, B., Oldfield, F., Richardson, K., Schellnhuber, H.-J. and Turner, B.L. 2004. *Global change and the Earth System: A planet under pressure.* Berlin, Heidelberg, New York: Springer Science & Business Media.

Sterling, S. 2001. *Sustainable education: Re-visioning learning and change. Schumacher briefings.* Schumacher UK, CREATE Environment Centre, Seaton Road, Bristol, BS1 6XN, England.

Sterling, S. 2014. At variance with reality: How to re-think our thinking. *Journal of Sustainability Education, 6.* Retrieved 30 January 2020 from http://www .jsedimensions.org/wordpress/content/at-variance-with-reality-how-to-re-think-our -thinking_2014_06/

Stubbs, W. and Cocklin, C. (2008). Teaching sustainability to business students: Shifting mindsets. *International Journal of Sustainability in Higher Education, 9*(3), 206–221.

Ungureanu, P. and Bertolotti, F. 2018. Building and breaching boundaries at once: An exploration of how management academics and practitioners perform boundary work in executive classrooms. *Academy of Management Learning & Education, 17*(4), 425–452.

United Nations. n/d. *Sustainable Development Goals.* Retrieved 30 January 2020 from https://sustainabledevelopment.un.org/sdgs

15. Does business ethics always have to be reactive?

Mark Heuer

INTRODUCTION

As academics, we are trained into an institutionalized mindset in which incremental advances are rewarded more than innovative theories and strategies. If we step outside this institutional mindset of incrementalism and dare to present a brave new theory, most of us experience the harsh realities of the reviewers' pens.

This reinforced aversion to risk taking often carries over to teaching, and even our personal lives, as the inertia of research interests, reputation building, and publishing can encourage following a path of least resistance. Specifically, if our research efforts are not rewarded for pursuing new and innovative ideas, it is also logical that we will focus our teaching on established environmental and social sustainability literature. There are many fine business ethics textbooks replete with case studies on familiar topics such as Enron's fraud, Monsanto's GMO practices, and Walmart's labor abuses. These are tried and true cases and can be taught effectively.

In defense of mainstream business ethics (which could also be referred to as reactive because we are dealing with what is in the past in an effort to project toward the future), it is much more likely that students will be familiar with sustainability issues that have become more settled and familiar. For example, in the business ethics classes I taught in 2010, I asked how many students were familiar with fracking. Less than 10 percent of the class, comprising undergraduate business seniors, were aware of the term, even though my university is within 30 miles of natural gas drilling in the Marcellus Shale Region of Pennsylvania. So, as a teacher wanting to engage students, do I choose to explain fracking and try to get them interested, or do I talk about the recycling projects on campus, opine about how few plastic bottles get recycled, and show some shocking pictures of plastic in the Pacific Ocean?

I passed on the opportunity to engage students on natural gas fracking in 2010. I made the excuse to myself that I did not have tenure and it was vitally important to get good student evaluations. Otherwise, I could be out of a job. The end result is that I ceded the opportunity to get students and myself involved in fracking in the early stages of the oil and gas industry's stealth strategy to develop fracking operations in the Marcellus Shale Region. In fact, the market entrance strategy of oil and gas firms into Pennsylvania from Texas and Oklahoma was so stealthy that fracked water, with its toxic substances, was dumped into sewage treatment plants in Pennsylvania, incapable of treating the toxic substances, before discharging into the Susquehanna River and its tributaries. This occurred even though the Susquehanna River provides drinking water for about two million people.

SHOULD I KNOW BETTER?

Yes, I should know better. I worked in Corporate America for twenty years managing federal contracts, US Capitol Hill for five years, and I also had a stint in a nonprofit producing items mandated by federal legislation. So, I cannot plead ignorance about how the game is played. The simple fact of the matter is that Corporate America is good at identifying profitable opportunities and seizing upon them quickly lest another competitor beat them to the punch. My job in Corporate America was to build a government contracting group by being the first mover in office products sales to the federal government. It was a hypercompetitive environment with double-digit growth rates and constant pressure to outbid the competition with service innovations and new products, such as 100 percent recycled copier paper or subcontracting efforts with minority-owned or women-owned firms.

On the other hand, my observation is that advocates for environmental and social sustainability, including nonprofits and academia, are not competing to be first movers by ringing the alarm bells and taking action at the earliest stages of an environmental or social sustainability issue.

With fracking, even the Sierra Club missed the fracking danger by taking $25 million from the fracking industry between 2007 and 2010 (Walsh, 2012). Apparently, they drank the PR "kool aid" that natural gas fracking was a bridge fuel from coal to renewables until the Sierra Club belatedly made an about face. Meanwhile, natural gas fracking grew rapidly in the Marcellus Shale Region, on a pace for Pennsylvania to become the second-largest producer of natural gas in the US behind only Texas.

With fracking well ensconced, I finally became personally involved with the fracking issue by attending a meeting at a local high school in 2011 led by two activists who were experiencing the direct effects of fracking on their property. I signed up to become involved and a group of us formed the "Riverfront Coalition for Clean Air and Clean Water" in Selinsgrove, Pennsylvania. I also joined another local nonprofit called "Shale Justice", which advocated more direct, high profile activism. Both organizations were made up of predominantly older activists from the 1960s and most were women. Aside from offering internships, we could never attract younger people. We did succeed in educating the public about fracking to counterbalance the barrage of pro-fracking commercials on TV and the giving of gifts, such as new firetrucks and equipment for sports teams in the Marcellus Region. We also succeeded in forming alliances with other nonprofits in the Marcellus Shale Region, including nonprofits in New York State who convinced Governor Cuomo to place a moratorium on fracking in New York. I also participated in the Horizontal Slickwater Hydraulic Fracturing Task Force of the Upper Susquehanna Synod ELCA Assembly (Evangelical Lutheran Church in America).

FRACKING RESEARCH

In research published in 2014 (Heuer and Lee) and 2017 (Heuer and Yan), it was clearly shown that respondents favored health and safety over economic security. This included private sector stakeholders. The 2014 paper showed that "overall, respondents prioritized the four categories in the survey as follows: Health & Safety (rank = 1); Communities (rank = 2); Economic Opportunity (rank = 3); Energy Security (rank = 4)" (p. 39).

Similarly, the 2017 paper found that, of the four categories used to organize the survey, i.e. economic opportunity, health and safety, communities, and energy security, the category of health and safety ranked first, as in 2012. Also, in the 2017 paper, female and more highly educated stakeholders favored a moratorium on fracking, while males and lower-income stakeholders supported fracking. I asked my students via a Blackboard Discussion Board why they thought women supported a moratorium on fracking more than men. This question piqued student interest in fracking and sustainability, as a whole.

AN OUNCE OF PREVENTION IS WORTH A POUND OF CURE

Through outreach from the nonprofit groups I joined, we learned through a variety of interactions that all of our state legislators were implacably supportive of the natural gas fracking industry and remain so today. This is the case even though, in addition to New York Governor Cuomo's moratorium on natural gas fracking, Governor Hogan, a Maryland Republican, signed into law a prohibition on fracking. In fact, there have been several recent defeats of natural gas pipelines in the eastern US, and the overproduction of natural gas that has led to many bankruptcies of oil and gas companies. However, in Pennsylvania this has been overshadowed by legislation approved by the Pennsylvania House and Senate to provide ongoing subsidies for "cracker" methane plants in Pennsylvania. These subsidies will basically provide a life-line to natural gas fracking in the Marcellus Shale Region. Governor Tom Wolf, a Democrat who campaigned on stronger regulation of fracking, signed the legislation into law.

The entire fracking experience in Pennsylvania brings to mind the old adage, "An ounce of prevention is worth a pound of cure." This gets at the core of the issue about business ethics needing to be more proactive, more innovative, and more willing to take risks. If we are more vigilant about being proactive rather than lobbying and protesting after the fact, we can be much more effective in supporting business ethics issues.

So how do we find the ounce of prevention? Basically, there are often hints about a major problem for a sustainability issue in its formative stage, if we pay attention. The Corporate America PR machine alerts us. Lobbyists get to work on politicians and we see efforts to shape public policy. For example, we started hearing a lot about how natural gas was the clean fuel that could provide the transition to renewable energy. In Congress, the Halliburton rule was passed that largely exempted federal environmental regulations from fracking. It could hardly have been clearer! As the institutionalization of the natural gas fracking industry in Pennsylvania was ignored, later efforts to fight fracking were dwarfed by the established industry infrastructure.

In the oil and gas industry, there is a predictable lifecycle. In the beginning stages of fracking in Pennsylvania, it was the smaller single rig operators who pioneered fracking. These operators were undercapitalized so that if they failed to strike a profitable pocket of gas in the first year or two, they were out of business. What these small operators found, however, was that, with horizontal drilling, their chances of striking it rich were much better than with traditional vertical drilling. So, this attracted the mid-sized regional drillers and soon they were hiring experts from Texas and Oklahoma to join in. By that time, an industry-friendly candidate for governor received generous campaign donations from Texas and Oklahoma. He became governor, drilling permits became easier to obtain, and, predictably, the state budget for gas drilling site inspectors and permit reviews fell.

So, it was too late. From that point on, environmental groups were in a reactive mode. At Shale Justice, our primary objective was to achieve a mor-atorium. Our belief was that if we got involved in dealing with the regulatory process (which is what many lawsuits do when you get right down to it), we were bound to win a battle here and there, which we did, but lose the war. Shale Justice disbanded when moratorium efforts failed. Now, with the financial problems of drillers, there is likely to be consolidation to establish scale and we will see Shell and Exxon become dominant. They can wait out the down cycle. The fight to block pipelines is really the last stand because once the pipelines are completed, liquefied natural gas plants, such as the one in Maryland, will become much more active in supporting markets abroad. Prices for natural gas abroad will be much higher and, as supply becomes tighter, gas prices in the US will rise as utility plants complete their conversion to natural gas from coal.

However, environmental activists did get focused on the pipeline aspect of fracking, which followed on the heels of the fracking boom. The numerous cancellations and/or postponements of gas pipelines is an indication that environmental groups at the local, regional, national, and global level have matured in strategizing about fracking.

Ironically, the pro-pipeline stakeholders incited the opposition by upsetting numerous anti-pipeline stakeholder groups, including Native American rights groups, environmentalists, property owners, and fishing and hunting groups. The networking necessary to block pipelines reminds me of the Starik and Rands (1995) AMR article, "Weaving an Integrated Web: Multilevel and Multisystem Perspectives of Ecologically Sustainable Organizations". While Starik and Rands (1995) pre-dates social media, it provides an excellent conceptual model. Granted, the pipeline stakeholders also provided lots of motivation to keep fracking opponents engaged. In addition to the historically lax regulation of pipelines, there are several other unpopular stakeholders to choose from. For example, FERC (Federal Energy Regulatory Commission) has a near-perfect record for approving pipeline permits. More locally, pipeline companies such as Williams have trained community organizers to appear at community events with carefully managed talk tracks, often leaving inquisitive citizens feeling disrespected. There is also social media about pristine forest land and trout streams to be plundered by compressor stations and other equipment needed to support the pipelines. Even Native American groups, not often visible in Pennsylvania, have protested the violation of ancient burial grounds by the Atlantic Coast pipeline, which Dominion Energy has canceled.

We do not know how the fracking issue will evolve. Right now, the price of natural gas is very low because of overproduction and many of the industry players are struggling with debt. The lifeline provided to the fracking industry by Pennsylvania politicians suggests a willingness to keep the industry alive.

The key point to always keep in mind through the long evolution of fracking is UN Sustainable Development Goal #7: clean, affordable energy. Natural gas is cyclical. The price in 2008 was several times what it is today. Also, natural gas is a fossil fuel and it is not a renewable source. The price and availability of solar power should be much more stable, not to mention clean. These are the key points I reinforce with my students. This generation of college students has known about climate change since elementary school. My role is to connect this awareness to specifics they relate to. For example, I told my students last semester that I was still driving my 2003 Honda hybrid. It impressed them that I had been driving a hybrid almost as long as they had been alive. In the next class, I asked how many thought they would drive an electric vehicle within the next ten years. Nearly all raised their hands.

TEXTILES: THE SECOND MOST POLLUTING INDUSTRY BEHIND FOSSIL FUELS

Unlike natural gas fracking, my students are very aware of fast fashion. When I ask how many students have been in an H&M store, most raise their hand. Moreover, social media has made fast fashion retailers and their "cheap chic" styles, just off the runway, ubiquitous.

However, similar to fracking, for the garment industry, which is my global research focus, there were hints of unsustainable environmental, labor, and safety aspects for a long time. For example, there were many reports about the textile industry moving from Northeastern mill towns to the nonunion Southern states such as North and South Carolina. It was not long before even the Carolinas were no longer competitive and the manufacturing moved to the Pacific Rim. When prices for garments and other textiles were sold in Walmart for prices the equivalent or less from decades earlier, it should have been obvious that lax labor, safety, and environmental regulations were probably a part of the mix.

We also began hearing about labor abuses in Indonesian factories making Nike products. Again, a PR blitz seemed to calm public concerns. I reenact the following PR mantra for my students as an educational technique:

> Nike provided better pay and working conditions in Indonesia than most other employers and Nike did not own the factories anyway, so there was not much they could do. And they have even sent a team of experts to fact find about the situation at the factories. Besides, the Nike factories are providing jobs to people who probably would not have a job otherwise. What is more, the US had labor issues like this when we were a developing country and look at us now. This is a normal part of growing into a developed economy.

I then ask students if they think Indonesian workers making Nike shoes earn a living wage. Then I play a YouTube video, *Nike Sweatshops: Are Nike's Workers Paid a Living Wage?* (Team Sweat, 2012). Phil Knight, founder and chairman of Nike, states that they are… absolutely! The YouTube video then takes the viewer on a walking tour to price out the daily cost of housing, food, water, and other basic essentials. Then the daily cost of living is tallied up against the wages earned. Most students come away believing the workers are not paid a living wage.

Phil Knight strongly believes Indonesian workers making Nike shoes earn a living wage even after the negative publicity Nike had received years prior to the YouTube filming. I became convinced, especially after a trip to India, that global supply chains powered the "race to the bottom" and Western consumers were the engine to its growth. During trips to Indonesia in the mid- and late 1990s, it became quite evident that Indonesia's labor situation was nothing like the US in its early years. The globalization of the economy and the disconnect between the factories in Indonesia and the retailers in developed countries is so pronounced, that it is difficult to compare with a Western context, where at least the rule of law has always been in effect, albeit imperfectly.

In an attempt to make some sense of the social and labor issues in fast fashion, I spent a summer studying supply chain literature. I was impressed by how little governance there was by host countries, as well as by developed countries, such as the US and in Europe. From this I concluded that there were very few barriers blocking lax enforcement of environmental, social, and labor standards, even though there were numerous voluntary global standards.

Then, on the morning of April 24, 2013, the eight-story Rana Plaza garment factory collapsed in Dhaka, Bangladesh, killing 1,134 and injuring more than 2,000. Most of the dead and injured were young women. As a result, I visited Asia in 2014 to learn more about fast fashion from Ms. Lizette Smook, a veteran of the garment manufacturing industry in Bangladesh and other Asian countries. We published a case study entitled *Remembering the Rana Plaza Workers: Change or Status Quo?* (Heuer and Smook, 2016), which, along with a teaching note, became a focus of my teaching and research. In this same period, I also published a book chapter on the global supply chain of the garment industry in a book entitled *Implementing Triple Bottom Line Sustainability into Global Supply Chains* (Heuer, 2016).

As I began teaching on global supply chains and the garment industry, I felt a need to at least try to present possible solutions for fast fashion. While doing summer research at the University of Kassel with Prof. Dr. Stefan Seuring, a leading scholar on sustainable supply chains, I discussed the idea of a book about consumer behavior to address fast fashion with a PhD student, Carolin Becker-Leifhold. Ultimately, we coauthored a book entitled *Eco-Friendly and Fair: Fast Fashion and Consumer Behavior* (Heuer and Becker-Leifhold, 2018), which contained chapters on sustainable consumption methods to avoid clothing going into landfills. My business ethics students have by now caught onto the trend of clothing rentals and sharing, so this topic is popular and engaging.

In one class, I had students develop and implement a semester project on assisting second-hand clothing stores with a marketing plan. It started off very well but tailed off as the semester wore on. (I had the same experience with sustainable food and farming projects, so I am still looking for a successful model for community projects such as this. Discussing ideas about sustainable fashion seem to be more interesting than the implementation!)

While I had published research on various aspects of the supply chains of the garment industry, a gap remained in terms of possible solutions at the factory level. I met with Mr. Avedis Seferian, CEO of WRAP (Worldwide Responsible Accredited Production), which is the largest third-party certification program of garment factories for addressing compliance with labor, safety, and environmental standards of host countries. Through the certification efforts of WRAP, garment factories in developing countries such as Indonesia qualify for garment orders with retailers in the US. I visited garment factories in Indonesia in early 2020 and interviewed managers regarding the WRAP compliance program. Subsequently, I published a case with Sage entitled *Third-Party Certification of Garment Factories: Antidote for the Race to the Bottom?* (Heuer, 2020).

Here again, an ounce of prevention is worth a pound of cure, as the success of the WRAP program is that it provides for ongoing audits to retain certification at an individual factory. The incentive for the factory is that certification makes them eligible for orders from US retailers, which pay higher prices than can be fetched in Asia.

Mr. Seferian appeared in my business ethics class to discuss the garment industry and the WRAP program. The interview was conducted on Zoom, with each student responsible for written questions, which they then posed to Mr. Seferian. For the students, it was the first they had heard about certification programs as a way to address environmental, social, and labor standards. This was interesting to me because my students had expressed support for sustainable fashion; however, their lack of awareness about certification programs suggested they were not aware about how sustainability was implemented.

CONCLUDING THOUGHTS

I focus on teaching my business ethics classes about fracking and fast fashion as the fossil fuel industry and textile industry are, respectively, the first and second most polluting industries globally. I have environmental lobbyists visit class and the CEO of WRAP speak via Zoom about the labor conditions in unregulated garment factories, but how do we avoid being reactive? How do we engage students to become involved?

Can we business ethicists somehow focus on the formative stages of a developing sustainability issue? I am not sure. We will not be able to change the incrementalism embedded in academic research. I see two possibilities. First, we can rediscover some of the concepts and techniques from the formative years of sustainability research. In particular, I see three concepts that can be useful in developing an early warning system, and then a first line of defense, in slowing the progress of a sustainability problem, while more permanent strategies can be implemented. Specifically, I am thinking, first, of applying environmental scanning to identify issues. Then, second, a plotting of the potential lifecycle of an issue to match it up with the lifecycle development of a comparable mature problem in order to gauge its severity along a probable progression. Lastly, third, it is a matter of developing the appropriate network to attack the issue before the PR firms and lobbyists put it out of reach.

Given the low cost of communication and the availability of technology to track issues globally, these three concepts are much more effective and affordable for identifying and strategizing on sustainability issues in the early, formative stages than previously.

Second, how do we go about managing a sustained effort, given that the reward structure of academia and the pocketbooks of Corporate America are not likely to oblige this?

For guidance and direction, the UN SDGs offer a holistic, yet cogent, set of principles to help chart and maintain a sense of direction given the distractions toward more micro-level concerns that confront us daily.

All of the Sustainable Development Goals (SDGs) represent their own maelstrom of sustainable challenges. To keep the challenges, once identified, before us, we can take each UN SDG and identify the three most important formative issues confronting each one. From that, we can prioritize them in terms of those we can impact most effectively. Then, we can develop a strategy and an implementation plan with timeframes for achieving interim goals. This is the process corporations go through in building their growth plan each year. Except here, we have a *set of principles* in the form of the UN Sustainable Development Goals (SDGs) to act as guideposts.

Principles are a bit like business strategy: if you develop them but then put them on the shelf and do not implement them, they maintain their high-mindedness, but little else. As an example of implementing principles, we can draw on the Sullivan Principles. These were developed in 1977 by a Philadelphia minister and sought to address apartheid in South Africa by advocating economic sanctions by corporations. Ultimately, the consistent, ongoing implementation of the Sullivan Principles led to growing concerns about apartheid, which was abolished ultimately (Perry Bullard Papers, 2020). The success of the Sullivan Principles depended on committed leadership and sustained efforts to do the hard work of lobbying corporations to, first, raise awareness of apartheid (thousands of miles away in South Africa) and, second, to convince corporate CEOs to persuade political leaders to pressure South Africa to end apartheid. While the fight to end apartheid in South Africa may now seem to be in the distant past, I recall the activism of my colleagues and me at the Lutheran School of Theology at Chicago (LSTC), where we con-structed a South Africa-style hut and conducted a sit-in to oppose apartheid in the late 1970s. We did this in the formative stage of the anti-apartheid effort, which for this reason was effective in educating a morally sensitive, but largely unaware, group of influential people.

For me, this sit-in effort awakened an awareness about environmental and social issues, as well as the importance of being proactive. I would not have had the drive to push for the solar energy plant at my university if I had not had the examples of leadership at LSTC with our anti-apartheid efforts. So, in my teaching I have taken steps to be more proactive as well in hopes of developing ethical leaders. One approach I have employed in the global business ethics class I teach is to have students decide collectively what topics they want to emphasize during the semester. I have developed the following options to be general enough to allow for a focus on current and developing issues:

Rank the following by the most (1) to the least important (9) ethical issues:

1. Fraudulent accounting or investment practices (such as the Wall Street meltdown in 2008).___
2. Failure to follow environmental regulations (illegal dumping/emissions or falsifying permits).___
3. Deceptive advertising or pricing (misrepresenting product attributes or pricing).___
4. Electronic privacy issues (government collection of phone records or business accessing credit and financial information without permission).___
5. Domestic or international labor and safe working environments (right to a living wage and safe working conditions).___
6. Discriminatory hiring and promotion practices and harassment in the workplace (race, gender, age, etc.).___
7. Campaign contributions to politicians by corporations or labor unions.___
8. Corporate governance (the requirement of government to treat shareholders as the first priority of public corporations).___
9. Government budget priorities (education, taxes, regulation, income support such as social security, Medicare, unemployment benefits, defense, law enforcement).___

No, business ethics does not have to be reactive! It is easier to be reactive for all the reasons discussed in this chapter. Being proactive requires taking risks in research, teaching, and service. Those who chose to make sustainability a focus in business schools in the early 1990s were often marginalized and faced the risk of not getting tenure. Now, business sustainability research is a vibrant field with several quality journal outlets. There are few, if any, incentives to be proactive, which is precisely why it is so important to inspire others in the same way my colleagues at LSTC did with the anti-apartheid protest.

REFERENCES

Heuer, M. (2016). Value Chain Connectedness as a Framework for Sustainability Governance. In L. Bals & W.L. Tate (Eds.), *Implementing Triple Bottom Line Sustainability into Global Supply Chains*. Sheffield, UK: Greenleaf Publishing, accessed June 14, 2021 at https://www.researchgate.net/publication/301348957 _Implementing_Triple_Bottom_Line_Sustainability_into_Global_Supply_Chains.

Heuer, M. (2020). *Third-Party Certification of Garment Factories: Antidote for the Race to the Bottom?* Thousand Oaks, CA: Sage Publications.

Heuer, M. & Becker-Leifhold, C. (2018). *Eco-Friendly and Fair: Fast Fashion and Consumer Behavior*. Abingdon, UK: Routledge.

Heuer, M. & Lee, Z.C. (2014). Marcellus Shale Development and the Susquehanna River: An Exploratory Analysis of Cross-Sector Attitudes on Natural Gas Hydraulic Fracturing. *Organization & Environment*, 27(1): 25–42.

Heuer, M. & Smook, L. (2016). *Remembering the Rana Plaza Workers: Change or Status Quo?* Ann Arbor, MI: William Davidson Institute, University of Michigan.

Heuer, M. & Yan, S. (2017). Marcellus Shale Fracking and Susquehanna River Stakeholder Attitudes: A Five-Year Update. *Sustainability*, 9, 1713, doi:10.3390/ su9101713.

Perry Bullard Papers. (2020). General Motors in South Africa: Secret Contingency Plan in the Event of "Civil Unrest", accessed July 21, 2020 at https://michiganintheworld .history.lsa.umich.edu/antiapartheid/items/show/201

Starik, M. & Rands, G. (1995). Weaving an Integrated Web: Multilevel and Multisystem Perspectives of Ecologically Sustainable Organizations. *Academy of Management Review*, 20(4), https://doi.org/10.5465/amr.1995.9512280025

Team Sweat. (2012). *Nike Sweatshops: Are Nike's Workers Paid a Living Wage?* YouTube.com.

Walsh, B. (2012). How the Sierra Club Took Millions From the Natural Gas Industry and Why They Stopped. *TIME*, February 2.

16. Students in action: faculty encouraging outreach and involvement

Gary Cocke, Joanna Gentsch, William E. Hefley, and Carolyn Reichert

While working with students in class and during student activities, faculty noticed gaps in students' understanding and implementation of the United Nations Sustainable Development Goals (SDGs). Supported by university and faculty initiatives, a new university-wide undergraduate elective course was developed to focus on the SDGs and community engagement. We examine the initiatives that led to the new course, including the personal faculty objectives that motivated the creation process. After the literature review, personal sustainability, and course structure, we conclude with our main results and areas for additional research.

COURSE INITIATIVES

Although students can be excited about sustainability, they often have a narrow focus on specific areas of interest. For example, they may include clean water but exclude hunger in considering sustainability. Students are often overwhelmed by the scope of some problems, and they may need guidance and information to empower them as agents of positive change (O'Flaherty & Liddy, 2017). A new elective, Sustainable Development Goals and Local Action, addresses these gaps. Available to all undergraduate students, it exposes them to the breadth and intersectionality of sustainability through the 17 UN SDGs and provides opportunities for application via service. The community-engagement learning class is supported by a university system grant and endorsed by Faculty Senate and the Sustainability Committee. Course enrollment is limited to foster the interactions important for deep connections and learning. Designed as a university course and housed in Undergraduate Studies, it first launched in fall 2019. The course enrollment has increased every semester and encouraged the creation of other service-based learning courses.

Over years of working with students on sustainability projects in co-curricular settings, the Sustainability Director observed the desire of students across disciplines to implement SDGs in practice, but students often thought of sustainability leadership as separate from their education. The challenge was to mentor students so they are empowered to lead change, while also helping them to understand that sustainability leadership should be an application of their education. The Sustainable Development Goals provide a framework, a context, and a language for sustainability where students can see the application of their education towards pressing environmental, social, and economic issues. The SDGs often represent an expanded definition of the scope of sustainability for students. The expansion allows them to see the interconnected links and broader applications.

We investigate the impact of off-campus engagement and personal experience in education for sustainable development (ESD). Course requirements shifted during COVID-19 to allow personal choice in community engagement. Since data are still being analyzed, this paper focuses on anecdotal results and personal experiences of students, faculty, and community partners. More rigorous research will follow in a later report.

LITERATURE REVIEW

On ESD, UNESCO (2020) advises the use of interactive, project-based, learner-centered andragogy. They advocate incorporating ESD throughout the entire learning environment "to enable learners to live what they learn and learn what they live."

Numerous papers document the benefits of experiential and interactive learning in sustainable development (Shephard et al., 2015; O'Flaherty & Liddy, 2017). Students gain competencies through holistic engagement, interdisciplinary perspectives, and learner-centered opportunities (Boeve-de Pauw et al., 2015). This encourages sustainable action and communication, empowering students to develop their skills to actively apply and promote SDGs (Shephard et al., 2015). Karatzoglou (2013) advocates for a curriculum that engages local partners to improve the effectiveness of ESD. Adapting sustainability to local themes and priorities gives additional meaning to the curriculum (Laurie et al., 2016).

Knowledge does not guarantee a change in attitude, motivation, or behavior (Kopnina & Meijers, 2014). Behavioral changes are often inconvenient and require effort. Arbuthnott (2009) writes that some of the strongest influences on behavior – concreteness and specificity – are often missing in ESD programs. Programs should help students translate intention into action, so students understand that they can control the behavior and their efforts will make a difference.

Students engaging in self-reflection using a simple resource audit were likely to modify their behavior (Savageau, 2013; Kopnina & Meijers, 2014).

Self-reflection, community engagement, and student interaction can improve engagement and behavioral change. Behavioral contagion, the tendency to repeat behavior exhibited by others, affects environmental and socio-economic issues, encouraging students to act in concert for a common goal (Frank, 2020). Students may over-emphasize advice on difficult tasks (Gino & Moore, 2006), giving self-reflection and guidance even more import. Personal values often drive sustainability engagement (Mulà et al., 2017), and faculty use personal examples and experiential learning to empower students (Cotton et al., 2007).

Administrative support is critical for effective ESD. The university organization and leadership must support sustainability initiatives (Harpe & Thomas, 2009). Lack of time, funds, and administrative support hamper efforts to incorporate sustainability into higher education (Barber et al., 2014; Moore, 2005; Laurie et al., 2016). Experimental course designs and pedagogy can face resistance from administration and colleagues. While Cotton et al. (2007) find high administrative support for sustainable development, this is tempered by studies (Barber et al., 2014) showing a lack of resources and support.

This course takes a holistic approach to ESD, combining active learning and community engagement to encourage student action on SDGs. Regional connections across a range of SDGs, enriched by personal experiences of faculty, community partners and students, enhance the student experience and encourage action and behavioral change.

PERSONAL SUSTAINABILITY

The coauthors cover a wide range of sustainability research, teaching, and service activities. Our course support includes instruction and design, securing grant funding and administrative backing from Faculty Senate and the Sustainability Committee.

The instructor coauthor has an academic background in ecology and conservation, and so his professional career in sustainability began through his passion for the environmental dimension of sustainability. As the field of sustainability evolved and frameworks like the SDGs increasingly embraced economic and social dimensions to build more holistic sustainability definitions, his passions related to the sustainability field also evolved. Still a passionate environmentalist, he views environmental issues as manifestations of social and economic sustainability concerns addressed in the SDGs. His pursuit of sustainable activities in his personal life are mostly challenged by limits of accessibility or resources. He sees that these factors limit action for all individuals and limit progress towards sustainability for systems as well. SDGs provide the language to identify these specific obstacles and allow students to connect individual experience to systemic challenges for sustainable development.

In his personal life, sustainability is the guiding force for most decisions he makes, and he loves to find the topic that resonates for a particular student. His personal sustainability actions include net-zero rooftop solar, energy efficiency throughout his home, native pollinator gardens, Bokashi composting for food waste, backyard chickens, organic vegetable gardening, organic land care, conscious consumerism and minimalism, cloth diapering, non-profit leadership, and donations to charities and social causes. He enjoys being in nature, working with his hands, and discussing his sustainability interests with others while hearing about their interests. He finds that most people have hobbies and interests related to sustainability, even if they did not realize the relation, and enjoys discussing how personal action for sustainability contributes to larger progress. He believes that personal conversations can break preconceptions that people may have regarding sustainability to make the field more inviting and accessible. These beliefs inspired him to codesign and teach an interactive course on SDGs.

His personal engagement with sustainability is shared in the classroom, across campus, and in the community through speaking engagements and personal actions. He created a Sustainability Service Honors Program and improved campus participation in numerous recognition programs (AASHE STARS, Bee Campus USA, Tree Campus Higher Education). He collaborated with a partner school to form a local Regional Center of Expertise (RCE) recognized by United Nations University in February 2019. Communication of sustainability has expanded through a newsletter, an annual report, and use of social media. These activities expand sustainability awareness and highlight opportunities for personal sustainability engagement. He incorporates these personal and professional experiences into the course instruction, showing students how to live a more sustainable life. This happens through guest lectures, personal examples, and reflections throughout the course.

The grant coauthor has a background in developmental psychology and extensive experience in community-engaged learning for undergraduates (Delano-Oriaran et al., 2015). She believes that collaborative relationships among students, instructors, and the greater community embed students' education in a broader context outside of the traditional classroom setting.

Through personal efforts and professional and community involvement, she attempts to translate SDGs into action as well as model and advocate for them with her students, friends, and family. She donates time and money to philanthropic causes focused on eliminating poverty, including initiatives to reduce unintended pregnancy, efforts to support children and families, and the campus food bank. She champions quality education and gender equality in her private life, often speaking publicly on these topics. She engages in positive psychology at home, living the examples from her class. Her family is intentional about food waste and shop and eat locally. They have installed a variety of devices to help reduce and manage water usage and installed energy efficient windows and fixtures. For the past four years, she has driven a fully electric car and lives close to work to reduce the commute. She is involved in political organizations that focus on social justice issues and actively devotes time and money to groups focused on reducing gun violence. She recruits qualified staff, many from under-represented groups, to facilitate community-engaged classes, providing quality work and compensation opportunities.

Her personal views motivate the creation of opportunities across campus for students and faculty to explore their unique interests in the SDGs. This led to the creation of service-learning classes that focus on aging, narrative storytelling in a hospice setting, immigrant communities, food security, health disparities, positive psychology, and gender inequality. She created a community-engaged class to address the unique challenges of college students with young children, an often-overlooked group. She shares these activities both inside and outside of the university, encouraging others to support their local community. Based on her personal and professional insights, she secured a university system grant for community-learning courses. The grants provide faculty an opportunity to develop courses that build on their personal sustainability interests and create a connected community of committed and engaged students, thereby enhancing academic and social belonging. In code-signing this course, she incorporated her personal contacts and importance of SDGs to individuals and society.

The Senate coauthor has published research and given presentations on corporate social responsibility (CSR) in sourcing, green IT, health and well-being, and accessibility in education and employment. He is involved with the university's Accessibility Committee and the Society for Disability Studies. He gained in-depth exposure to corporate social responsibility (CSR) and sustainability practices working with an India-headquartered global IT firm (Radice & Hefley, 2005). Through interviews with thousands of staff, he learned first hand of their involvement with volunteerism through programs such as adult literacy and Tata Consultancy Services Maitree (Jinnia, 2014), and Sevalaya (2020). His research culminated in an introduction to social responsibility and sustainability in the outsourcing industry that highlighted select companies and their actions undertaken (Babin and Hefley, 2013). Many firms were aspirants for the IAOP/ISG Global Outsourcing Social Responsibility Impact Award (GOSRIA) (IAOP, 2019). He helped establish this award to recognize outsourcing service provider excellence in CSR activities that foster community, workplace training, communication, environment, and giving. He served as a member of the US technical advisory group on the development of ISO guidance on social responsibility (ISO, 2010).

On a personal level, his family buys local farm products from local farmers or a subscription box. They have served as a drop-off and distribution point for other subscribers. They recycle and repair rather than replace whenever possible. They switched one vehicle from a high-power V8 engine to a hybrid auto. He supports organizations such as Sevalaya to overcome barriers from poverty, hunger, and lack of access to education. He has made a personal commitment to high-quality education, leaving the consulting world to return to the classroom full-time. He makes decisions about his research and interests, such as smart cities, driven by the SDGs.

His work in outsourcing and service science (Hefley & Murphy, 2008) led him to incorporate CSR into his teaching, research, and service activities. A strong advocate for corporate social responsibility, he develops courses on managerial approaches to environmental management systems, social responsibility guidance, and stakeholder concerns on firm environmental, social, and governance performance. He emphasizes real-world managerial experience and addresses sustainability leadership as an application of management education, incorporating his personal experiences through discussion and development of case studies. Students learn sustainable business management practices, the role of the private sector in global sustainability development and environmental initiatives, and progress towards the 17 UN SDGs. His support of initiatives on and off campus has improved accessibility, awareness, and engagement in CSR activities by firms and individuals. He provides administrative support for this course, promoting its development through Faculty Senate and on-campus committees.

The final coauthor chaired the Sustainability Committee. Her first introduction to sustainability was environmental, but she now integrates social and economic aspects too. Her active research includes climate change and governance. She incorporated SDGs into an executive education course, encouraging business leaders to include sustainability in their decision-making criteria.

She donates to the local foodbank through various community initiatives. She minimizes food waste, buys locally, and selects organic and non-toxic alternatives. Through xeriscaping and native plant selection, she reduces irrigation use. Her family recycles, repairs, reuses, repurposes, or donates. She prefers paperless choices to reduce waste. During COVID-19 restrictions, her family learned about other cultures through food, music, geography, and history. She volunteers at a local nature center. Kanashiro et al. (2020) highlight additional areas she could incorporate, generating ideas for further involvement.

While she encourages sustainability, public and private responses are mixed. Some push for more action, while others see costs to minimize. The average view is positive, so she continues to share her experiences. She offers the course administrative support, promoting development through the university-wide Sustainability Committee.

The coauthors' focus ranges from environmental to socio-economic aspects, bringing a range of perspectives to the course development and covering all the UN SDGs. As highlighted above, we communicate these personal actions and connect them to our teaching, research, and service. Verbal communication occurs at least weekly with written communication bi-weekly to students, faculty, staff, and the community, depending on the topic. Each author is available to support the course administratively and academically as needed. The semester length allows us to typically focus on one SDG per week.

COURSE STRUCTURE AND SUPPORT

The college experience can overwhelm undergraduates. The scope of the UN SDGs is vast and complex. This can lead to inertia in student action. A framework can guide and encourage students to act on sustainability initiatives. Starting small and encouraging student engagement increases the likelihood of sustainable actions.

When faculty are involved with a particular topic, students respond to their excitement. Modeled behaviors and personal stories influence student enthusiasm while conveying values and expectations. There is a large body of literature highlighting the role of narrative literature and behavior, including behavioral contagion in organizations (Martin, 2016). If that is the case, then faculty action and examples will resonate even more.

To change student perceptions of sustainability and invite new voices to the conversation, this means authenticity is the most important tool for faculty. The authors sincerely believe that SDGs represent the challenges that this generation of students must solve, and they believe that every student can leverage their education and interests towards addressing these issues. Student engagement with SDGs is never prescriptive and is unique according to the interests and talents of the individual. Sincere interest in each student and their perspectives, a shared belief in the goals, demonstrated subject matter expertise, and exhibited personal sustainability behaviors demonstrate authenticity, build credibility, and inspire action.

The Sustainable Development Goals and Local Action course is a junior-level class available to all undergraduates. It can be used as a free elective for any undergraduate degree and satisfies the community service requirement contained in some degrees. It has extensive administrative support, building on faculty personal sustainability to encourage student involvement with SDGs and the community. The course is designed to deepen student knowledge of sustainability by defining the field through the 17 UN SDGs, to build breadth across the range of goals and to resolve gaps in understanding and action. Students learn that sustainability issues are interconnected and interdependent, and addressing SDGs requires contributions from many disciplines and perspectives. There are weekly reflections and frequent guest lectures from community partners. Many of the guests are drawn from the authors' personal contacts and provide real-world context for the SDGs. A service-learning project, models of systemic change, and a mock climate negotiation add to the interactions.

Sustainability is personal for the instructor coauthor, and he aims to make it personal for each student. When given the opportunity to design a course teaching sustainability, he designed the class using the 17 UN SDGs as a framework to define the field and connect personal student interests to sustainability. During the course, students explore the SDGs as they connect their personal experiences to local efforts and see how these efforts contribute to international progress on SDGs. Students learn where their interests lie and are encouraged to value the input and expertise of others through discussions. They develop greater respect for others' perspectives as they collaborate to address issues of local importance with a community partner. As faculty and community partners share their personal and professional experiences, students build their personal experience with the field of sustainability. This deepens their understanding and empowers them to apply their education and skills in that direction.

To address sustainable leadership, students focus on a community partner. For the first two semesters, this partner was identified by a faculty call for projects through the area's RCE. In fall 2019, students developed a food waste plan for their community partner, a local school district. Along with discussions on campus food waste, students received a guest lecture from the regional Environmental Protection Agency office about food waste. For spring 2020, students examined how SDGs correlate with root causes of hunger, partnering with a local food bank. Students received guest lectures on childhood poverty and from the food bank staff.

Fall 2020 required remote instruction due to COVID-19. The virtual modality made guest lectures easier to schedule. Since students identified individual community partners based on personal interest, guest lectures did not need to focus on building knowledge towards a single deliverable, and instructional time was freed from collaborative project planning. This provided additional class time for local experts, who provided contextually relevant examples of how sustainability issues manifest locally, and the relevant work done to address these issues.

RESULTS

The course expands student engagement and a self-reported sense of purpose. It supports students' connection to the UN SDGs in their community, and faculty communicate both their personal interests and broader sustainability goals. Using information from course evaluations and collected as part of the university system grant, we will analyze student satisfaction and learning of SDGs.

Based on preliminary assessment data, students demonstrated self-reported changes in belonging and mattering via their emerging role as agents of social change. Students reported enhanced cognitive insight, empathy and perspective taking, and a greater awareness of important global issues and social justice concerns. As co-creators of their learning experience, students honed their leadership and professional skills, feeling empowered and finding meaning because of their sustainability efforts. They actively contributed to their peers' development and shaped their own interpersonal growth. This is particularly impactful as participating students were more likely than their peers to be transfers, first-generation, or under-represented minorities.

A sense of purpose has been shown to enhance student engagement (Xerri et al., 2018). Student engagement, academically and with campus and community, is important for student retention and development (Meer et al., 2018). Students are enthusiastic about service-learning classes with connections to the SDGs. Their community participation can affect meaningful changes. One example was the class response to COVID-19. As the food bank's work became more crucial, the class responded by providing more immediate assistance. They created and promoted a social media campaign to help the food bank provide for families in need. The class, in conjunction with the Office of Sustainability, raised $3,335 for the community through their digital campaign. Their engagement was detailed in an article circulated extensively on and off campus.

In fall 2020, the intersectionality of sustainability issues was brought into sharper focus. The pandemic made inequalities more apparent as suffering disproportionately affected people of color and low-income communities. The racial justice movement provided additional urgency. Discussions deepened and abstract sustainability issues were suddenly made very tangible and immediate for students. This context helped students understand that addressing social and economic issues is imperative to make progress on environmental issues. Faculty shared their personal experiences, adding authenticity and encouraging student activity across the UN SDGs.

While the match between personal and professional sustainability interests is not one-to-one, we are committed to maintaining or increasing our sustainability support both personally and professionally and communicate that support publicly through words and actions. Our course empowers students to act on the UN SDGs and build community on and off campus. During the course, we share our own actions for sustainability and hopes for the future, and help students develop an awareness of, and demonstrate, sustainability leadership through their own actions.

REFERENCES

Arbuthnott, K. D. (2009). Education for sustainable development beyond attitude change. *International Journal of Sustainability in Higher Education*, *10*(2), 152–163. https://doi.org/10.1108/14676370910945954

Babin, R., & Hefley, B. (2013). *Outsourcing professionals' guide to corporate responsibility.* 's-Hertogenbosch, Netherlands: Van Haren.

Barber, N. A., Wilson, F., Venkatachalam, V., Cleaves, S. M., & Garnham, J. (2014). Integrating sustainability into business curricula: University of New Hampshire case study. *International Journal of Sustainability in Higher Education*, *15*(4), 473–493. https://doi.org/10.1108/ijshe-06-2013-0068

Boeve-de Pauw, J., Gericke, N., Olsson, D., & Berglund, T. (2015). The effectiveness of education for sustainable development. *Sustainability*, *7*, 15693–15717. https://doi.org/10.3390/su71115693

Cotton, D. R., Warren, M. F., Maiboroda, O., & Bailey, I. (2007). Sustainable development, higher education and pedagogy: A study of lecturers' beliefs and attitudes. *Environmental Education Research*, *13*(5), 579–597. https://doi.org/10.1080/13504620701659061

Delano-Oriaran, O., Penick-Parks, M. W., & Fondrie, S. (2015). *The Sage sourcebook of service-learning and civic engagement.* Thousand Oaks, CA: SAGE.

Frank, R. H. (2020). *Under the influence: Putting peer pressure to work.* Princeton, NJ: Princeton University Press.

Gino, F., & Moore, D. A. (2006). Effects of task difficulty on use of advice. *Journal of Behavioral Decision Making*, *20*(1), 21–35. https://doi.org/10.1002/bdm.539

Harpe, B. D., & Thomas, I. (2009). Curriculum change in universities. *Journal of Education for Sustainable Development*, *3*(1), 75–85. https://doi.org/10.1177/097340820900300115

Hefley, B., & Murphy, W. (2008). *Service science, management and engineering: Education for the 21st century.* New York: Springer Science & Business Media.

IAOP. (2019). The Award: The IAOP/ISG Global Outsourcing Social Responsibility Impact Award (GOSRIA). Accessed on June 4, 2021 at https://www.iaop.org/content/19/165/3427/default.aspx

ISO. (2010). *ISO 26000:2010 Guidance on Social Responsibility.* ISO. Accessed on June 4, 2021 at https://www.iso.org/iso-26000-social-responsibility.html

Jinnia. (2014). Corporate Social Responsibility: Case of TCS. *Journal of Indian Research*, *2*(1), 96–102.

Kanashiro, P., Rands, G., & Starik, M. (2020). Walking the sustainability talk: If not us, who? If not now, when? *Journal of Management Education*, *44*(6), 683–698. https://doi.org/10.1177/1052562920937423

Karatzoglou, B. (2013). An in-depth literature review of the evolving roles and contributions of universities to education for sustainable development. *Journal of Cleaner Production*, *49*, 44–53. https://doi.org/10.1016/j.jclepro.2012.07.043

Kopnina, H., & Meijers, F. (2014). Education for sustainable development (ESD). *International Journal of Sustainability in Higher Education*, *15*(2), 188–207. https://doi.org/10.1108/ijshe-07-2012-0059

Laurie, R., Nonoyama-Tarumi, Y., Mckeown, R., & Hopkins, C. (2016). Contributions of education for sustainable development (ESD) to quality education: A synthesis of research. *Journal of Education for Sustainable Development*, *10*(2), 226–242. https://doi.org/10.1177/0973408216661442

Martin, S. R. (2016). Stories about values and valuable stories: A field experiment of the power of narratives to shape newcomers' actions. *Academy of Management Journal, 59*(5), 1707–1724. https://doi.org/10.5465/amj.2014.0061

Meer, J., Scott, S., & Pratt, K. (2018). First semester academic performance: The importance of early indicators of non-engagement. *Student Success, 9*(4), 1–13.

Moore, J. (2005). Barriers and pathways to creating sustainability education programs: Policy, rhetoric and reality. *Environmental Education Research, 11*(5), 537–555. https://doi.org/10.1080/13504620500169692

Mulà, I., Tilbury, D., Ryan, A., Mader, M., Dlouhá, J., Mader, C., Benayas, J., Dlouhý, J., & Alba, D. (2017). Catalysing change in higher education for sustainable development. *International Journal of Sustainability in Higher Education, 18*(5), 798–820. https://doi.org/10.1108/ijshe-03-2017-0043

O'Flaherty, J., & Liddy, M. (2017). The impact of development education and education for sustainable development interventions: A synthesis of the research. *Environmental Education Research, 24*(7), 1031–1049. https://doi.org/10.1080/13504622.2017.1392484

Radice, R., & Hefley, W. E. (2005). *Interpreting SCAMPISM for a People CMM® Appraisal at Tata Consultancy Services* (CMU/SEI-2005-SR-001). Software Engineering Institute. https://resources.sei.cmu.edu/library/asset-view.cfm?assetid=7309

Savageau, A. E. (2013). Let's get personal: Making sustainability tangible to students. *International Journal of Sustainability in Higher Education, 14*(1), 15–24. https://doi.org/10.1108/14676371311288921

Sevalaya. (2020). A Temple of Service [video]. YouTube. Accessed June 4, 2021 at https://youtu.be/C-z9i7QnbpM

Shephard, K., Harraway, J., Lovelock, B., Mirosa, M., Skeaff, S., Slooten, L., Strack, M., Furnari, M., Jowett, T., & Deaker, L. (2015). Seeking learning outcomes appropriate for 'education for sustainable development' and for higher education. *Assessment & Evaluation in Higher Education, 40*(6), 855–866. https://doi.org/10.1080/02602938.2015.1009871

United Nations Educational, Scientific and Cultural Organization (UNESCO). (2020). *Education for sustainable development: A roadmap.* UNESCO. Accessed on June 4, 2021 at https://unesdoc.unesco.org/ark:/48223/pf0000374802

Xerri, M. J., Radford, K., & Shacklock, K. (2018). Student engagement in academic activities: A social support perspective. *Higher Education, 75*(4), 589–605.

17. Student sustainability knowledge gained from classroom and field experience

Dave Nelson and George Ionescu

INTRODUCTION

In today's global world, disruptions continuously create barriers to a company's sustainability (Boisot, 1995). Historically, resources were abundant and readily available for businesses with access to them, like lumber for residential development. However, that changed with increasing resource scarcity, reductions in accurate forecasting and impediments for long-term planning. In the late 1960s scientists discovered mounting evidence of global warming with adverse climate changes from industrialization, like massive glaciers melting in the Arctic and Antarctic. More recently, around the beginning of 2019, global communities were caught unaware when a novel coronavirus evolved into a Pandemic (COVID-19), even though scientists had previously investigated the possibility of such a virus in the 1970s (Weiss, 2020).

There seems to be an overarching need reflecting 'New Normal' adaptation, not only for adjusting to climate change (Tong and Ebi, 2019), but also mitigating growth of viral epidemics like coronavirus, creating disruptions for business sustainability and global populations. It is important to understand that climate change initiatives and morphological variations (adaptations) of coronavirus could continuously impact existing and future generations. To adapt to the effects of global warming and mitigate coronavirus, research and logistics have become an integral part of survival for businesses and populations around the world.

Education is a key factor in order to foresee potential issues and facilitate the necessary changes, requiring knowledgeable faculty for mentoring students throughout their education, preparing them to be responsible citizens and caretakers of the environment. The world as we have known it now requires continuous research and adaptation. Strategies to accommodate these necessitate input from all stakeholders for appropriate pathways, achieving social and environmental harmony for sustainable societies.

According to Cordero et al. (2020), there is a growing literature base in climate change education, embedding more math and science in core curricula. Even though knowledge of these is helpful for understanding global warming, not everyone has been able to accomplish that level. However, the paths of Corporate Social Responsibility (CSR) and Circular Economy (CE) can help both businesses and communities bridge this knowledge gap, enabling them to understand how to accommodate shifts in climate change while working towards sustainability (Agudelo et al., 2019; Allen and Craig, 2016; Morton et al., 2017).

COURSE DESIGN, PROCESSES AND OUTCOMES

This course was designed to enhance student knowledge, embedding sustainability, climate change and the Pandemic, enabling them to share information both in class and with local communities during meetings with them. When students provide these talks, their potential for sustainable decision-making is enhanced, along with a holistic approach for mitigation and adaptation. Conveying course information helps facilitate a proactive approach towards social and environmental harmony, facilitating mitigation and adaptation strategies for sustainability.

The Overview of Logic (Figure 17.1) is continuously discussed during class, helping students visualize logic and pathway development, reflecting learning and proactive strategies to mitigate disruptions and achieve sustainability. For Course Design, Activities encompass work performed by students helping to educate the community. Focus houses all course material including CSR/CE, climate change initiatives and the novel coronavirus. Processes reflect community engagement and education, literature search, report writing, and examples from faculty real-world experience. Lastly, Outcomes illustrate how knowledge gained during the course is used holistically for proactive decision-making, enabling students to work towards a more sustainable society.

FACULTY IMPACT ON STUDENT BEHAVIOR

When faculty are committed to working on environmental and social issues for adapting to climate change and mitigating adverse effects of coronavirus, they often demonstrate these commitments, both personally and professionally (Kanashiro et al., 2020a). In doing so, instructors act as working examples where all individual actions matter, when reflecting on environmental and social sustainability. When the instructor provides examples of personal and

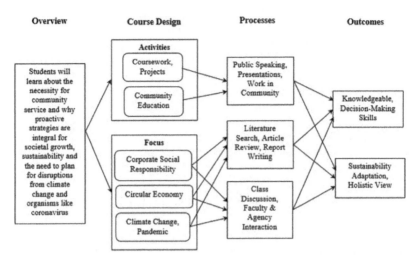

Source: Adapted from Cordero et al. (2020).

Figure 17.1   *Overview of logic and development for course designed to*
                 help students become forward-thinking, proactive citizens
                 seeking sustainability with social and environmental harmony

professional work in Corporate Social Responsibility and Circular Economy, some students discover they have similar experiences and share them during class. The Overview of Logic model was selected for the course because of consistency with United Nations Sustainable Development Goals and continuously reflected on by both the students and instructor.

While the Overview of Logic (Figure 17.1) is discussed in depth throughout the semester, the Biggs' 3P Model Logic (Figure 17.2) is utilized by the instructor as the framework for course progression, which is periodically reflected on during class with students, illustrating conceptual parallels with the Overview Logic.

The first factor (Presage) involves both faculty and student experience, where student focus reflects on epistemic beliefs and previous experience (professional and academic), along with their values and interests. Faculty focus also involves epistemic beliefs, expertise level and their professional affiliation, as well as skills and curricula design aptitude. The other factors (Process and Product) encompass student learning performance and outcomes, stemming from positive faculty influences, based on their depth of engagement. When students perform research and become knowledgeable, they can then 'walk the talk', like the faculty members.

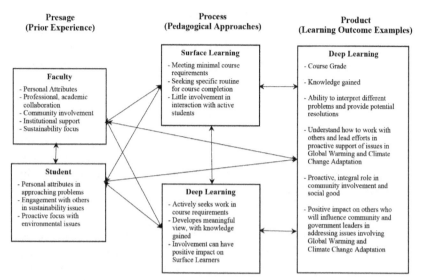

| Presage (Prior Experience) | Process (Pedagogical Approaches) | Product (Learning Outcome Examples) |

Presage
(Prior Experience)

Process
(Pedagogical Approaches)

Product
(Learning Outcome Examples)

Surface Learning
- Meeting minimal course requirements
- Seeking specific routine for course completion
- Little involvement in interaction with active students

Deep Learning
- Course Grade
- Knowledge gained
- Ability to interpret different problems and provide potential resolutions
- Understand how to work with others and lead efforts in proactive support of issues in Global Warming and Climate Change Adaptation
- Proactive, integral role in community involvement and social good
- Positive impact on others who will influence community and government leaders in addressing issues involving Global Warming and Climate Change Adaptation

Faculty
- Personal Attributes
- Professional, academic collaboration
- Community involvement
- Institutional support
- Sustainability focus

Student
- Personal attributes in approaching problems
- Engagement with others in sustainability issues
- Proactive focus with environmental issues

Deep Learning
- Actively seeks work in course requirements
- Developes meaningful view, with knowledge gained
- Involvement can have positive impact on Surface Learners

Source: Adapted from Kanashiro et al. (2020b).

Figure 17.2   Biggs' 3P Model Logic

PRESAGE (FACULTY AND STUDENT EXPERIENCE)

The first element houses the instructor's prior experience and personal characteristics from their professional and academic skill set development. The student's presage factors include personal attributions, engagement, professional and academic experience, values, interests and demographics. When both instructor and student possess these characteristics, there will be much more engagement within the classroom. One of the main reasons for this is that active students more easily understand when the instructor is walking the sustainability talk.

One of the authors worked in the field of Environmental Science where industrial sites were routinely audited by state and federal agencies. The focus of his work was to verify sites were in compliance with safe work practices. During class lectures, the logistical progression of site evaluation is discussed, along with hazard reduction strategies. Students familiar with these issues and knowledgeable about environmental compliance often become actively involved with the lecture material and class discussions, enhancing not only their retention of information, but also for the class as a whole.

One of the authors consults with organizations on the east and west coasts of the US, providing insight in the areas of climate change, carbon footprint reduction and sustainability. The need for this is illustrated during class with discussions of social and environmental community outreach efforts. During these talks, growth of Corporate Social Responsibility dimensions during the Industrial Revolutions are outlined. Stories are shared about his company reaching out into the community to provide free training courses and certification for unemployed individuals, along with employment assistance after course completion.

After establishing the need for CSR, focusing on product design and development, the benefits of Circular Economy are discussed where waste is treated as a resource. This runs countercurrent to the historical effort for handling waste in a Linear (disposable) Economy, with marginal effort in recycling and more landfill waste. During these lectures, one of the authors discusses shifting towards Circular Economy, with an example where he repurposed barn redwood for a display case in his home. He also talks about his family seeking firms that manufacture net carbon-neutral products from recycled material, like Hewlett-Packard. Further discussions reflect benefits from better product design, lengthened life cycles and component remanufacture, shifting linear supply chain logic to circular (Bakker et al., 2014).

Students also learn about how the family of one of the authors strives for zero waste – recycling/repurposing everything they can, along with taking steps to reduce their carbon emissions. In tandem with this is the discussion of alternative energy, decreasing dependency on petroleum products and limiting travel by plane.

PROCESSES (SURFACE AND DEEP LEARNING)

This element houses student attitude reflecting two learning types – Surface and Deep. These stem from both the instructor's presage influence and the student's level of engagement with course material and experiential learning, understanding benefits, utility of mentoring and working on projects together (Filho et al., 2016). Surface is indicative of minimal student effort to complete coursework, with little intention of using information subsequent to course completion. Deep students exhibit strong engagement and comprehension, with good potential for positively impacting surface learners. Their efforts enable them to achieve a good conceptual understanding of applications when presented with predicaments after course completion. Understandably, this approach is very helpful when CSR and CE are involved because of the necessity to reflect on the dimensions of economic, social and environmental issues (Kanashiro et al., 2020b). When this happens, the student learns to walk the sustainability talk.

To help clarify misconceptions about climate change and novel coronavirus, one of the authors' developed progressive sets of Discussion Board Questions requiring student input throughout the semester. Everyone was tasked to discuss their adaptation and mitigation strategies after reading course material on climate change, and a booklet on the novel virus (Nelson and Seavey, 2020) was written by the author and former student, describing effects of the coronavirus and mitigation strategies, professionally and in their personal lives. Excerpts from both are shown in Appendix A. The Questions offered a good learning experience where requirements involved interactive discussions among all students in the class encompassing both their personal lifestyle and professional work environment.

PRODUCT (LEARNING OUTCOMES)

The final element primarily applies to student Deep Learning Outcomes, based on their knowledge, intuition, initiative and ability. Working student examples are found in their attending meetings, performing research and fieldwork, along with community involvement with sustainability issues reflecting Corporate Social Responsibility and Circular Economy. One of the authors and a student developed the Pandemic booklet noted above to help demystify coronavirus and mitigation logistics. They also teamed up with previous students from two of his courses who were seeking work with environmental issues in the community. They met with local businesses to discuss means/methods of virus mitigation and global warming adaptation, with follow-up phone conversations with further information. Currently, the author and a student are working on another booklet titled 'Introduction to Global Warming', offering adaptation strategies for climate change issues to help forward-thinking individuals, along with and small and large businesses.

SURVEY INSTRUMENTS

Over time, different methods were explored for student input on course improvement. Initially, end-of-term surveys were used, enabling one of the authors to progressively improve the course. This continued until the effects of climate change and coronavirus acted synergistically, adversely impacting student learning and knowledge retention. To evaluate the effect and help students understand the need for changes, the authors developed two surveys reflecting Biggs' 3P approach, commencing in the fall of 2020. The first survey occurred during course startup before any material was distributed. The second took place at the end of the semester, after completion of all course work. Representative items are shown for surveys in Appendix B.

Both surveys reflect course material on supply chain sustainability, Corporate Social Responsibility, Circular Economy and the focal firm's relationship with suppliers and customers, with items on adaption to global warming and mitigation of the coronavirus. Integral to this are both the Brundtland Commission's recommendation for enhancing growth potential of future generations and the United Nations Sustainable Development Goals (2015).

The first (Pre-Course Survey) houses Presage, evaluating student knowledge level about supply chain issues, global warming, novel coronavirus, Corporate Social Responsibility and Circular Economy issues. The second (Post-Course Survey) reflects the Process element, containing more comprehensive issues – UNSDGs, Brundtland Commission, sustainability and supply chain adjustments for mitigating the Pandemic and adapting to global warming. Both surveys were developed to capture student interest in relevant topics for optimizing course quality and content, enabling them to become more adept in adapting to shifting work environments in a globally changing world.

SURVEY RESULTS

Pre-Course Survey results showed students were knowledgeable about the novel coronavirus, but the instructor needs to clearly explain that when the existing novel coronavirus is terminated, other organisms may surface, also requiring mitigation. Students understood supply chain importance, but more information should be included to improve understanding chain utility, especially in the context of climate change, pandemic and post-pandemic recovery. They also understood proactive focal firms are key players with their corporate culture influencing the entire chain, addressing environmental and social issues. While suppliers are integral for supply chain survival, students need to understand that using a few select ones is optimal. More material should be included regarding the importance of supplier influence on customer relationships, along with the need for focal firms sharing information with competitors.

Post-Course Survey results exhibited good comprehension of course material, along with positive statements (final questions) reflecting the Pandemic Overview booklet. Several students stated they will personally and professionally use the booklet. Preliminary conclusions indicate there are more students who will use knowledge gained during the course in their future lives, as compared to previous courses.

CONCLUSIONS

This chapter houses personal and professional experience of instructors as examples for mentoring/teaching students about the necessity of adaption to global warming and mitigation of coronavirus. By the term's end, there is a commonality, linking the instructor with some of the students. When students realize they have a good level of knowledge, enabling them to help with tasks like the United Nations Sustainable Development Goals, they become better citizens and caretakers of the planet we all live on.

Corporate Social Responsibility and Circular Economy are discussed throughout the course, with integration into the culture of forward-thinking firms. This is important for student learning because their knowledge can help reduce the gap between historical paths of business (linear economies, abundant waste) versus firms utilizing forward-thinking, Circular logic, benefiting the society with reductions of waste and pollution.

There are two models. The first (Overview of Logic) displays student progression, enabling them to continually visualize their progress. The second (Biggs' 3P) reflects how their experience, learning objectives and outcomes parallel the first model. The Biggs' 3P Model is an important framework for the instructor to stay on track with during course progression, which is integral for student learning.

A key benefit of this course allows for adaptation over time to better accommodate global changes, help mitigate new coronavirus strains and address supply chain disruptions. The benefit of utilizing Biggs' 3P is shown in post-survey results illustrating the difference between previous course surveys (without use of Biggs' 3P), exhibiting a lower rate of student awareness and commitment for proactive changes (4–5%), compared to results of this course with a higher rate (8–10%). Notably, the course should improve more over time since this is the first effort.

REFERENCES

Agudelo, M. A. L., Jóhannsdóttir, L. & Davídsdóttir, B. (2019). A literature review of the history and evolution of corporate social responsibility. *International Journal of Corporate Social Responsibility*, *4*(1), 1–23.

Allen, M. W. & Craig, C. A. (2016). Rethinking corporate social responsibility in the age of climate change: A communication perspective. *International Journal of Social Responsibility*, *1*(1), 1–11.

Bakker, C., Wang, F., Huisman, J. & den Hollander, M. (2014). Products that go round: Exploring product life cycle extension through design. *Journal of Cleaner Production*, *69*, 10–16.

Boisot, M. (1995). Is your firm a creative destroyer? Competitive learning and knowledge flows in the technological strategies of firms. *Research Policy*, *24*, 489–506.

Cordero, E. C., Centeno, D. & Todd, A. M. (2020). The role of climate change education on individual lifetime carbon emissions. *PLOS ONE*, *15*(2), 1–23.

Filho, W. L., Sheil, C. & Paco, A. (2016). Implementing and operationalizing integrative approaches to sustainability in higher education: The role of project-oriented learning. *Journal of Cleaner Production*, *133*, 126–135.

Kanashiro, P., Rands, G. & Starik, M. (2020a). Walking the sustainability talk: If not us, who? If not now, when? *Journal of Management Education*, 1–30, https://www

.researchgate.net/publication/342539371_Walking_the_Sustainability_Talk_If_Not_Us_Who_If_Not_Now_When, accessed 14 June 2021.

Kanashiro, P., Iizuka, E. S., Sousa, C. & Ferreira Dias, S. E. (2020b). Sustainability in management education: A Biggs' 3P model application. *International Journal of Sustainability in Higher Education, 21*(4), 671–684.

Morton, S., Pencheon, D. & Squires, N. (2017). Sustainable development goals (SDGs) and their implementation. *British Medical Bulletin, 124,* 81–90.

Nelson, D. & Seavey, D. (2020). Overview of the COVID pandemic, https://drive.google.com/file/d/1B_qhwzoCXIX-365RPvho4_YIQNP7eJ-W/view, accessed 29 May 2020.

Tong, S. & Ebi, K. (2019). Preventing and mitigating health risks of climate change. *Environmental Research, 174,* 9–13.

UN Sustainable Development Goals (2015). https://www.undp.org/content/undp/en/home/sustainable-development-goals.html, accessed 6 June 2021.

Weiss, S. R. (2020), Forty years with coronaviruses. *Journal of Experimental Medicine, 217*(5), 1–4.

APPENDIX A. EXCERPTS FROM PANDEMIC AND GLOBAL WARMING BOOKLETS

Historical Perspective – Viruses evolved about 1.5 billion years ago, utilizing gene pools with an ongoing process of natural selection enabling them to survive in changing environments. Some viruses have high mutation rates allowing them to quickly adapt to environmental conditions. They also have much shorter life spans than humans, which results in a rate of adaptation that is comparatively accelerated to that of humans.

Evolution – Coronavirus 2019 (COVID-19) evolved as a respiratory disease caused by the SARS-Cov-2 virus. Origin of the virus is believed to be from China and has taken pandemic proportions encompassing all continents around the world.

Virus Spread – The virus has impacted all aspects of day-to-day operations in families and businesses, globally. Coronavirus is believed to be primarily spread between people in close contact with one another via respiratory droplets from infected people.

Symptomology – Warning signs (symptomatic) can occur in a few days, or up to two weeks after exposure. Health professionals believe roughly 40% of infected people do not display any symptoms, refer to them as asymptomatic. Infection of the virus causes COVID-19 resulting in illness (fever, cough, difficulty breathing), ranging from mild to severe.

Climate Change

Discovery – Svante Arrhenius discovered global warming and it was confirmed by Thomas Chamberlin. They stated that industrial and human activities warm earth due to carbon dioxide emissions. At the time it was thought human activity was too miniscule to make any real impact. They theorized the ocean took care of our emissions and acted as an effective carbon sink.

Industrial Revolution Emissions – There was a lag time between start of emissions and visible impact, likely due to lack of technology capabilities to measure CO_2 in the atmosphere. New technology occurred during WWII, when scientists discovered an increase of temperature due to greenhouse gases trapping heat in the atmosphere.

Emergence of Environmental Protection – The Environmental Protection Agency was established in the 1970s in the US, which led to more government regulations involved with pollution. The Clean Air Act of 1970 was enacted due to negligent practices by emissions of firms damaging natural habitats and the environment.

Atmospheric Levels of CO$_2$ – Levels of carbon dioxide in the atmosphere were found to be increasing over time, based on findings from the National Oceanic and Atmospheric Administration.

Kyoto Protocol – This Protocol houses the 'United Nations Framework Convention on Climate Change', whereby industrialized countries commit their efforts towards reduction of Greenhouse Gas Emissions in accordance with mutually stipulated targets.

APPENDIX B. PRE-SURVEY AND POST-SURVEY EXAMPLE ITEMS

Pre-Course and Post-Course Questionnaire Parallels

Pre-Course

Not all firms in the supply chain need to focus on ethics. Only the focal firm needs to do that.

I wanted to take a class involving business adaptation to COVID-19 and Global Warming because that knowledge will prepare me better for a career.

I am interested in supply chain sustainability and looking forward to learning about it in the course.

I have experience with adapting to Global Warming from a personal and business standpoint.

Companies with Corporate Social Responsibility in their culture will have a better chance for surviving supply chain disruptions.

Accommodations made by firms for coronavirus, will no longer be necessary when the virus is eliminated.

I have knowledge about helping to mitigating COVID-19 with PPE, masks and distancing, from a business standpoint.

Post-Course

The class offered insights into mitigation of the coronavirus and adaptation to global warming.

Discussion Questions added value to the course because they included current topics I may need to reflect on for my career.

Knowledge about effects of COVID-19 and Global Warming can help future decisions in my career and personal life.

In supply chains, ethics are as important in offshore operations as they are with the onshore suppliers and the focal firm.

Discussion Board Forums were beneficial, enabling us to discuss how we deal with COVID-19 in our personal life and at our job.

I did not know Additional Reading Material and interactions with other students would be required. If I had known that, I would have enrolled in a course only requiring reading from a text, with exams and quizzes.

Having an instructor with experience in real world issues covered during class added value to the course.

The course material was relevant for current issues faced in today's global economy.

I learned skills in the course to help in my career pursuit and about organizations that work in sustainability and supply chains in the global world.

The course helped me realize the importance of my degree and how I can use the information learned to help work towards the necessary changes.

PART IV

Faculty personal sustainability as social
movement

18. The power of faculty sustainability practices helping businesses drive social change: an interview with Jessica Yinka Thomas

Patricia Kanashiro

BUSINESSES WITH A PURPOSE: THE RISE OF CERTIFIED BENEFIT CORPORATIONS (B CORPS) AND BENEFIT CORPORATIONS AS A SOCIAL MOVEMENT

Since the 1970s, there has been a surge in corporations redefining their business purpose beyond maximizing shareholders' wealth. In particular, there is an emergence of hybrid organizations in the form of social enterprises that combine both economic and social welfare logics (Grimes, McMullen, Vogus, & Miller, 2013; Pache & Santos, 2012). Certified B Corporations and Benefit Corporations are some examples of popular forms of hybrid organizations.

The definitions of Certified B Corporations and Benefit Corporations often overlap, but these categories represent significant differences. Certified B Corporations (or B Corps) are businesses that meet the highest standards of verified social and environmental performance, public transparency, and legal accountability to balance profit and purpose. Businesses that seek a B Corp certification must complete a free assessment (B Impact Assessment) to measure how the company's operations and business model impact workers, communities, environment, and customers. Once the company meets the performance targets, the company amends its bylaws to formally include all stakeholders' interests as fiduciary duties of the board of directors (Kim, Karlesky, Myers, & Schifeling, 2016). B Corps need to pay an annual fee and be reassessed every three years to maintain certification. There are currently 3,500 companies certified as B Corps, representing 70 countries and 150 industries (Certified B Corporation, n.d.). Some examples of B Corps include Patagonia, AltSchool, King Arthur Flour, Sun Light & Power, and Natura.

Benefit Corporations differ from B Corps in at least three different ways. First, a Benefit Corporation is a legal business structure, similar to a Limited Liability Company (LLC) or a Corporation (Inc). A Benefit Corporation designation is attributed to a business with modified obligations to create public benefit, beyond generating profits to shareholders (B Lab, n.d.). Second, the legal classification as a Benefit Corporation is available only in 38 US states, including the District of Columbia. Finally, to obtain legal status as a Benefit Corporation, the organization provides a self-report on environmental and social performance. Such a report does not require third-party verification, nor must the organization be re-certified after a period.

Nonetheless, both Benefit Corporations and B Corps are leaders of a global social movement that emphasizes the role of business as a force for good. Businesses are called to redefine their identities and redesign business models to generate positive impacts for all stakeholders (Kim & Schifeling, 2016). The social movement phenomena generated attention among academics who became increasingly interested in further developing the topic through research and teaching. In 2016, several faculty members got together and launched the B Academics, a platform for academics to share ideas and foster collaboration. The next section presents a short bio of Professor Jessica Yinka Thomas, the founder and current leader of B Academics, followed by an interview that took place virtually, on July 23, 2020.

ABOUT JESSICA YINKA THOMAS

Professor Jessica Yinka Thomas was born in Miami and lived in Nigeria and Senegal because of her father's career in academia. She spent most of her life traveling between the two regions and became aware of the deep social and economic discrepancies and the finite nature of environmental resources. Her awareness has significantly informed her professional work and personal choices, as she always knew that she needed to bridge the gap between the two worlds (Cornell, 2020; Morrison, 2020).

Jessica holds a Bachelor's in Engineering degree from Stanford University and an MBA from Duke University. One of her first professional positions was as a product designer for LeapFrog, a toy company specializing in designing educational and technology-based toys to support children's learning (Taris & Stasio, 2020). Jessica was particularly excited when she received letters from parents reporting how LeapFrog toys helped their kids learn Science, Technology, Engineering, and Math (STEM) concepts. She then realized that businesses have a great potential to be a driver for good, especially considering businesses' power to grow and scale up impactful initiatives (We Are the Arc Benders, n.d.).

Figure 18.1   Photograph of Jessica Thomas

After leaving LeapFrog, Jessica worked as an independent consultant and frequently took travel breaks between projects to explore her creativity (Page, 2012). Inspired by her sabbatical experiences around the world, she wrote a novel entitled *How Not to Save the World* (Thomas, 2012) that narrates the adventures of Remi Austin, a frustrated fundraiser who turns to criminal activities to be able to continue raising funds for her nonprofit organization. She later launched the sequel *How Not to Make Friends* (Thomas, 2018). Jessica believes that storytelling can be a powerful tool to engage individuals to find their path to leaving the world a better place (Page, 2012).

Since 2007, Jessica held several positions in academia and currently serves as Assistant Professor of Practice (new title as of January 2021) and Director of the Business Sustainability Collaborative at the Poole College of Management at North Carolina State University (NC State University) in Raleigh, NC. As a faculty member, she believes that she can use her teaching to scale up the idea that businesses can be used as a force for good. In 2016, she received the Bill Clark Award in recognition of her leadership in creating and scaling up the NC State B Corp Clinic (Morrison, 2020) and leading the B Academics movement. The NC State B Corp Clinic has engaged over three hundred students from eight colleges working on nearly sixty projects for companies in North Carolina and around the world interested in becoming a Certified Benefit Corp. Students have the unique opportunity to work with real companies and help these organizations to improve their social and environmental strategies and impacts (NC State University, n.d.).

She also led the creation of B Academics, the global B Corp academic community, to advance the state of academic study into business as a force for good (B Academics, n.d.). B Academics was formally launched in 2016 during a roundtable meeting at The Wharton School of the University of Pennsylvania in Philadelphia. The network now includes almost 2,000 educators, researchers, and practitioners engaged in researching and teaching business as a force for good.

Jessica is a great advocate of continuous improvement and intentional living. She believes that both are key factors for successfully implementing sustainability practices at a personal level. She believes in the power of connecting like-minded people to exchange ideas, share solutions, and be supportive of one another to build a strong coalition for a systemic change; and she invites each of us to join the B Academics movement.

INTERVIEW WITH JESSICA YINKA THOMAS

Tell us about your career and how you started.

I have spent the better part of the last 15 years in academia working at Duke in the School of Engineering and at the University of North Carolina Kenan Flagler Business School, and more recently at the North Carolina State University in the Poole College of Management. During all of this time, my work has focused on the power of business to drive social and environmental impact. I have experience working in nonprofit organizations and as an engineer. My first career was designing interactive educational children's toys, but I believe the arc has been at the intersection of education and impact.

How can businesses have a positive impact?

For-profit businesses are powerful agents for change due to their potential for scale. Social and environmental impacts can be embedded in the business model and, therefore, in the way the company operates. A new business model is fundamental to a necessary systemic change. We are experiencing a powerful and visible social movement shifting the purpose of business from creating value for shareholders to creating value for all stakeholders more broadly defined. Over the last ten years, I have become very interested in the B Corps movement. I think the B Corps certification process offers a particularly powerful, comprehensive, and transparent model, and provides a roadmap that clearly defines the requirements for businesses to drive positive social and environmental impacts. Students – who are future business leaders across sectors – and business managers can easily understand and apply the certification model.

How was the B Academics organization launched?

The B Academics has existed for as long as the B Corps movement has existed. Still, we have just recently formalized this network of educators and researchers interested in studying and teaching about B Corps, engaging with the B Corps managers, and connecting with the broader community. I have to say that faculty have been interested in the topic for a long time, long before we came up with the idea of formally launching the B Academics.

In 2016, while I was a faculty member at North Carolina State University, my school colleague Rosanna Garcia and I started connecting with like-minded colleagues. In that year, the North Carolina Research Triangle (comprised of North Carolina State University, University of North Carolina at Chapel Hill, and Duke University) and the B Corp community were invited to host the B Corp Champions Retreat event in Durham, NC.

This is a side story, but, in that same year, the state of North Carolina passed a piece of legislation called HB2, referred to as the "bathroom bill" that, among many other things, required people to use public restrooms based on the gender assigned in their birth certificate. This bill was a very discriminatory piece of legislation when it comes to our LGBTQ community. As a result of that legislation being passed, the B Corp community decided not to host the B Corp Champions Retreat in North Carolina. The event ended up being held in Philadelphia, hosted by the Wharton School of the University of Pennsylvania.

The B Academics network has grown organically since 2016, when Rosanna Garcia, Joel Gehman, and I worked with Wharton to host a small gathering for the first annual B Academics Roundtable. We are now a global network of close to 2,000 educators, researchers, and practitioners from 52 countries representing over 600 different organizations. Since 2016, we have continued to host the annual B Academics Roundtable along with workshops, webinars, and other events designed to connect, engage, and inform our network. In August 2020, we welcomed researchers from around the world for a virtual paper development workshop entitled "Nuts and Bolts of Research on B Corps: Empirical Methods and Research Design." This paper development workshop was designed to help the research community define and push boundaries of B Corps research, especially as related to empirical methods and research design. Scholars from across North America, Latin America, Europe, Asia, and Africa have expressed interest in getting engaged in our network and we are excited to continue building out our partnerships, resources, and events as we launch a formal membership platform in 2021.

How do you engage and build relationships between businesses and students?

B Academics utilize a range of strategies for connecting students and businesses from guest lectures to case studies to experiential learning programs like the B Impact Teams. Faculty are developing innovative strategies to inspire and engage students in the B Corp movement.

For example, Mary Baldwin University has developed an entire MBA program organized around B Corps and the B Impact Assessment. B Academics is working to develop a catalogue of teaching resources including cases, articles, videos, and podcasts, and build a global community of educators to share best practices in teaching. Moreover, we have developed a teaching guide for the second edition of the *B Corp Handbook* [Honeyman & Jana, 2019]. The book *Better Business: How the B Corp Movement is Remaking Capitalism* [Marquis, 2020] also has a teaching guide associated with it.

In my undergraduate business ethics course, I weave a discussion of B Corps through the semester as we explore ethics from a stakeholder engagement perspective. I invite guest speakers like Lauren Ginsberg, a Poole College alumna and a general manager at Athleta, which is a certified B Corp and a subsidiary of Gap Corporations. We do a case on New Belgium Brewery, which is also certified as a B Corp, to study open book management and employee ownership and strategies to engage workers. At North Carolina State University, we host an annual B the Change speed networking event, where students have an opportunity to learn more about B Corps and connect with representatives from local B Corps for employment opportunities.

In my MBA Sustainable Business Strategy course, I use the second edition of the *B Corp Handbook* as one of the course texts and host guest lectures from B Corp leaders, and the students work in teams on a practicum project with aspiring and Certified B Corps to help strengthen the students' impact business models as part of our B Corp Clinic program.

Our B Corp Clinic at North Carolina State University is an example of a B Impact Team, an experiential learning model developed by B Lab to connect students to impact improvement consulting projects with local businesses. In addition to the NC State B Corp Clinic, the B Impact Teams model has been adapted and implemented by a number of faculty in the B Academics network, including Kristin Joys at University of Florida, Fiona Wilson at University of New Hampshire, Jake Mosely at University of Georgia, and others. Maria Ballesteros-Solas at California State University Channel Islands, is even working to develop a curriculum for middle- and high-school students to introduce them to B Corps.

How do your personal stories inform what you do daily at home and on a personal level? How do you leverage those experiences in the classroom and with other stakeholders?

I grew up in West Africa, both in Nigeria and in Senegal. I think I have a very different background and experience from many of my colleagues and many of the students I interact with. My experiences shaped how I feel about our world, including our access to natural resources. I remember when I was a kid, whenever we had any products, we would use every last drop of it. We never wasted anything, whether it was water or other natural resources. We had a periodic water outage, and we didn't have water for days. We also had power outages, and we didn't have power for days. And so, even today, I think of all of these precious and finite resources, and that's definitely influenced my view from an environmental perspective. Growing up in West Africa, I was so aware of the wealth disparity because so many people in West Africa don't have access to education, healthcare, and other basic human needs. So, I think the way I grew up has significantly informed the work that I have done.

I see my work as a way to communicate and educate my students, work with companies, and work with people in my community. I want to increase the awareness that we are interconnected with the environment and other people, and how those interactions look from a systems perspective. As a teacher, I have the opportunity to interact with hundreds of students every year, and I believe I play an important role in making them aware of these issues and the critical role that each of them can play.

As a consumer, I am very thoughtful about my purchase decisions. I live in a relatively small home compared to my peers, and our family has one car, which is fewer than families of similar size but certainly more than most families in the world. One of the most impactful decisions I made when it comes to living sustainably is to live in a small house. That decision has a multiplier effect from reduced energy use to purchasing fewer things to fill the house.

I also try to be thoughtful about small decisions, so I will look for that B Corp certification label when I am shopping. For example, I shared an experiment on social media called the B Corp Challenge to see how many B Corp products or services I could use in one day. I looked at the products that I had at my house, took note of where I went to lunch, where I was shopping, and the kind of services I bought. I was able to identify at least fifty different B Corps certified products and services that I could use throughout the day. When I do shop, I increasingly seek out B Corps. From the bed and pillow I woke up on, to the products I used to brush my teeth, wash my face and style my hair, to the food and drinks I consumed, the clothes I wore, my bank, the stores I shopped at, the platform I used for publishing my novels, I was reminded that you can find B Corps in almost every industry and walk of life. We also happen to have a wealth of B Corp leaders right here in North Carolina.

What are some of the challenges you have encountered, and how do you share those challenges with the various stakeholders?

I believe sustainability is something that you are never going to be perfect at practicing, but it is about improving everyday choices. It is about how we evaluate our options, how we teach our classes, and make purchase decisions. I will admit, not every single one of my purchases is sustainable. I believe it is more of a practice and being aware of your choices and decisions and every day waking up with an intention to be purposeful and to be thoughtful to do the best that you can on that day. We are all hopefully moving in the same direction with a shared vision. But we will never be perfect. So, you may never be a perfect vegetarian, but you can every day make progress towards that vision, maybe consuming less meat. And perhaps some days you are a perfect vegetarian and some days you are not but do not see that as a failure. Instead, look at it as a process and a practice and not even necessarily that you are always going forward because sometimes you may go backward.

It is more about awareness intention and making progress over the long term. So, it is less about failing or succeeding, but every day being a learning opportunity about yourself. It is also about pushing yourself to learn more about the world and to learn more about new companies, new products, and new ways of doing things, and taking that learning into the next day.

What is the connection between the B Corp movement and the United Nations Sustainable Development Goals (SDGs)?

The B Corp movement is about continuous improvement, which is one of the reasons why I connected with the movement. It requires companies to continue to assess, evaluate, and improve their impact. Moreover, the NC State Pool College of Management is part of the United Nations Principles for Responsible Management Education (PRME) network. We are committed to integrating the SDGs into our teaching, research, and engagement with business and the communities. The SDGs provide a tangible way for students to think about goals and impact, reflect how the goals are measured from different perspectives, and connect with the ideals. We can also examine how businesses are engaging with the SDGs, including corporate reporting and impacts. Business managers can evaluate how they engage with their supply chain, recruit employees, and design and sell products and services. From an aggregate perspective, students can understand the power of businesses when we pull together tens of thousands of global corporations, all working on different aspects of the SDGs, and the power that businesses can have.

What is your advice to other faculty members on how to increase their personal efforts and sustainability practices and how to communicate those efforts to students in the classroom and to other stakeholders?

Well, I would just say join the B Academics! We need to think about how we can bring together the faculty community to share best practices to share inspiration to identify opportunities for collaborative teaching and research. We are building a community of like-minded academics, and I think there actually has been a lot of sharing of both personal and academic perspectives on various topics.

We are building relationships that aren't just going to serve our academic careers, but that will help us as human beings as well. And, so, yes, my advice would be, join us. Be part of the conversation. I think everybody has something to share, and we can create an ongoing dialogue and space where academics share best practices, inspire each other, and engage with each other.

What are your future plans?

We are preparing to launch a formal membership for B Academics in January 2021. B Academics has been operating informally for several years. In 2019 we became a formal 501(c)3 nonprofit with a very engaged founding board of directors. A membership structure is a natural next step to better engage the faculty, student and industry representatives in our network. As part of the membership launch, we have created an online resource repository of hundreds of teaching, research and engagement resources to inform our network. We are starting to work on research translation series to inform academics, the B Corp community, and industry more broadly. Membership will give us a better sense of the existing teaching, research, and engagement work in our network and create an opportunity to share best practices, identify opportunities for research collaboration, and strengthen engagement.

REFERENCES

B Academics (n.d.). *Mission and Vision*. http://www.bacademics.org/bcorporation.net, accessed January 26, 2021.

B Lab (n.d.). What is a Benefit Corporation? https://benefitcorp.net/, accessed November 1, 2020.

Certified B Corporation (n.d.). bcorporation.net, accessed November 1, 2020.

Cornell, L. (2020). The wisdom of women: Jessica Yinka Thomas, Founder & President of B Academics. *Conscious Company*, January 22. https://consciouscompanymedia .com/social-entrepreneurship/the-wisdom-of-women-jessica-yinka-thomas-founder -president-of-b-academics/, accessed November 1, 2020.

Elkington, J. (1997). *Cannibals with Forks: The Triple Bottom Line of 21st-century Business*. Oxford: Capstone Publishing Ltd.

Grimes, M. G., McMullen, J. S., Vogus, T. J., & Miller, T. L. (2013). Studying the origins of social entrepreneurship: Compassion and the role of embedded agency.

Academy of Management Review, 38(3), 460–463. https://doi.org/10.5465/amr.2012 .0429

Honeyman, R. & Jana, T. (2019). *The B Corp Handbook: How You Can Use Business as Force for Good (2nd edition)*. Oakland, CA: Berrett-Koehler Publishers.

Kim, S. & Schifeling, T. (2016). Varied incumbent behaviors and mobilization for new organizational forms: The rise of triple-bottom-line business amid both corporate social responsibility and irresponsibility. *SSRN Electronic Journal*. http://dx.doi.org/ 10.2139/ssrn.2794335

Kim, S., Karlesky, M., Myers, C., & Schifeling, T. (2016). Why companies are becoming B Corporations. *Harvard Business Review*, June 17. https://hbr.org/2016/06/why -companies-are-becoming-b-corporations, accessed November 1, 2020.

Marquis, C. (2020). *Better Business: How the B Corp Movement Is Remaking Capitalism*. New Haven, CT and London: Yale University Press.

Morrison, S. (2020). Jessica Thomas featured as world-changing woman in business. *Poole College of Management News*, January 24. https://poole.ncsu.edu/news/2020/ 01/24/jessica-thomas-featured-as-one-of-13-world-changing-women-in-business/, accessed November 1, 2020.

NC State University (n.d.). Business Sustainability Collaborative. https://bsc.poole .ncsu.edu/b-corporations/b-corp-clinic/, accessed November 1, 2020.

Pache, A. C. & Santos, F. (2012). Inside the hybrid organization: Selective coupling as a response to competing institutional logics. *Academy of Management Journal*, 56(4), 972–1001. https://doi.org/10.5465/amj.2011.0405

Page, L. (2012). New author spotlight! Meet Jessica Yinka Thomas. *Habits & Novelties*, February 3. http://literarylegs.blogspot.com/2012/02/new-author-spotlight-meet -jessica-yinka.html, accessed November 1, 2020.

Taris, J. & Stasio, F. (2020). From designing toys to promoting business sustainability: Meet Jessica Yinka Thomas. *North Carolina Public Radio*, June 1. https:// www.wunc.org/post/designing-toys-promoting-business-sustainability-meet-jessica -yinka-thomas, accessed November 1, 2020.

Thomas, J. (2012). *How Not to Save the World (Remi Austin #1)*. Lulu.com

Thomas, J. (2018). *How Not to Make Friends*. Lulu.com

We Are the Arc Benders (n.d.). *"B" ing the Change: Connecting Students to Business as a Force for Good*. https://www.wearethearcbenders.com/b-ing-the-change/, accessed November 1, 2020.

19. From personal to professional: a reflective account of academics engaging with sustainability

Louise Obara, Te Klangboonkrong, Gary Chapman, and Regina Frank

INTRODUCTION

Sustainability has become part of everyday life for university academics. For instance, faculty aim to help businesses see the benefits of organisational sustainability (e.g. Hart & Ahuja, 1996) and prepare students for a sustainable future by embedding it in the curricula (Wu & Shen, 2016). These efforts can raise awareness of and address the sustainability crisis, and equip organisations and students with the skills and knowledge to operate sustainably. Higher Education (HE) has seen increasing pressure for action on sustainability, translating into the emergence of sustainability rankings (Times Higher Education, 2019); yet, critics often view sustainability-related activities as mere marketing opportunities for HE (Barth, 2013), primarily leading to ceremonial compliance (Meyer & Rowan, 1977), but without meaningful change.

Developing an effective sustainability agenda in universities has been challenging (Barth, 2013). A small, yet growing number of researchers highlight the importance of academics in driving sustainability (Barth & Rieckmann, 2012). The current approach has largely focused on selected university operations (e.g. energy savings), and curriculum design, with limited success (Menon & Suresh, 2020). Kanashiro et al. (2020) suggest that business school academics need to go further by assessing, sharing and increasing their own personal sustainability actions and behaviours, thus calling for further research in this area.

This chapter responds to this call by undertaking an autoethnographic study focused on the collective reflection on our individual perceptions of and behaviours on sustainability, as four academics working in a UK business school. We explored our personal perceptions, actions and motivations relating to sustainability and their overlap with our professional lives. In this process, we introduced the lead user perspective from innovation studies, which focuses on individuals who are 'ahead of the curve' by engaging significantly and early in problem-solving, expecting high benefits due to dissatisfaction, intrinsic and extrinsic motivation (von Hippel, 1986; Bilgram et al., 2008). Lead users differ from the early adopters often studied as they are typically the key drivers of innovative solutions to major problems (von Hippel, 2005), and provide significant input and influence prior to the solution launch (Hienerth & Lettl, 2017). Thus, examining lead users can be valuable to reveal innovative mindsets, and sustainability behaviours in the faculty, which can advance understanding and guide the behaviours of other academics toward a more sustainable future.

The academic debates on lead users can be divided into two streams of literature with different foci: the lead user method (e.g. Lilien et al., 2002) or lead user characteristics (e.g. Franke et al., 2006), with this chapter focusing on the latter. In our autoethnographic study, we exploited a compositional difference between the users, with half of the authoring team (or study participants) actively involved in contributing to the university and business school's sustainability agenda and activities (i.e. lead users in their professional activities), whilst the other half represent more regular business school academics (Hienerth & Lettl, 2017). By juxtaposing 'lead' and 'regular' users, we extend von Hippel's lead user concept, and reveal more nuanced insights on faculty sustainability behaviours. Our autoethnographic approach allowed us to collectively reflect on the qualitative data gathered during an in-depth group discussion on sustainability (Alvesson et al., 2008). All authors participated and probed each other's responses to increase the richness in our data. Thereby, we respond to Kanashiro et al.'s (2020) call for research into the kinds of personal sustainability practices that faculty engage in and how these vary by level of involvement in sustainability activities (e.g. teaching) in their professional role.

LEAD USERS AND THEIR ROLE IN EDUCATION FOR SUSTAINABLE DEVELOPMENT

Our perception of driving sustainability in universities is consistent with organisational innovation, which is defined as "the implementation of a new organisational method in the firm's business practices, workplace organisation or external relations" (OECD, 2015, p. 51). Our key takeaway is that, even with seemingly prescriptive practices, successful adoption depends on the organisational fit (Ansari et al., 2010), making off-the-shelf transferability almost impossible. Considering that the sustainability agenda represents a situation "where practices, causality, and performance are hard to understand and chart" (Wijen, 2014, p. 302), the intensity of this challenge multiplies.

As some contextualisation will be necessary (Ansari et al., 2014) for the success of university sustainability, organisational learning and social learning seem pertinent for generating locally relevant knowledge (Argyris & Schön, 1978). In this line of thinking, social interactions and shared mental models are crucial for a new practice to be tailored and adopted successfully in an organisation. Key to this is the assumption that knowledge is tacit and 'sticky' (von Hippel, 1986) as it resides within each individual and collectively between organisational members (Nonaka & Konno, 1998), which perpetuates distinction between organisations. Thus, for faculty sustainability behaviours to be impactful in improving sustainability, social learning will need to occur between organisational members. We argue that a key source of social learning, and in turn sustainability progress, will be lead users, identified as individuals who are 'ahead of the curve' and engage significantly in solving problems, such as sustainability, to identify acceptable solutions. Such individuals possess novel knowledge and experience, and develop more strategically important ideas (Lilien et al., 2002) that, when shared, can contribute to a valuable pool of locally relevant knowledge to drive sustainability.

In our context, lead users refer to individuals who have been significantly engaged in university and faculty sustainability activities and agendas before these became part of their official remit, and who aim to drive change and identify solutions to the crisis of sustainability in universities. This significant professional engagement is typically accompanied by engagement in their personal lives, where lead users independently engage to solve problems (Lettl et al., 2008). We argue this complementarity between the professional and personal engagement of lead users in solving problems can also be observed in the faculty lead users who have shown innovative behaviours that can advance other 'regular' academics' understanding and influence change in their behaviour. That these lead users/faculty members are part of the organisation is crucial to this effort, as unlike the traditionally discrete standalone innovation problems, the sustainability agenda is location-specific, and continuous social learning will be necessary to induce lasting and systemic change. To redress this theoretical deficit, we propose the following research questions: What personal and professional sustainability behaviours do academics engage in, and to what extent have the former influenced the latter? How does engagement and motivation for engagement in faculty sustainability behaviours differ between sustainability lead and non-lead user academics? What are the personal mindset characteristics and sustainability behaviours of lead users that can help advance the implementation of sustainability and the SDGs in higher education?

METHODOLOGY

To explore the sustainability actions of the four authors involved in this chapter, an autoethnographic approach was adopted. This method "seeks to describe and systematically analyze personal experience in order to understand cultural experience" (Ellis et al., 2011, p. 273). As with all ethnographic approaches, the goal is to develop an in-depth understanding of the social world (the phenomenon under study) by examining the construction and interpretation of that world by its participants. In our case, our experiences are the focal point of analysis (Rock, 2007). We specifically adopted an interactive interview for this study, in which "researchers and participants – one and the same – probe together about issues that transpire, in conversation, about particular topics" (Ellis et al., 2011, p. 273). This method is particularly suited to groups whereby a well-established relationship exists and, as a result, views are expressed openly and freely, and individuals can probe and challenge the views of others. This method was particularly suited for our study, as we have worked in the same department for four years, and for the past year collaborated on a research project hence have developed a good working relationship.

We agreed on a set of key questions for the "interactive interview" that reflected the call for contributions, and focused on our perceptions, actions and motivations for sustainability and how these came into our professional lives. The conversations occurred in summer 2020 and lasted for approximately three hours. One author served as facilitator to ensure all questions were discussed and everyone had an equal opportunity to express their views. The meeting was recorded and transcribed, and the data were analysed by two authors initially, who coded the data into themes, and shared this with the remaining authors. This initiated further discussions about the data and themes, with subsequent amendments made.

As summarised in Table 19.1, all study participants are senior lecturers, and between them teach and research various business-related topics, including sustainability. Engagement with sustainability in their personal and professional lives varies greatly between participants, with A and B self-identifying with the "lead user" role, having worked on sustainability long before it became a focus for business schools. Participants C and D viewed themselves as "non-lead" or "regular" users, meaning that they were aware of sustainability and embrace it partly in their personal lives, but this had not translated into their professional lives.

FINDINGS

The findings are presented in two parts. The first compares and contrasts the personal and professional sustainability actions of participants. The second explores some of the debates authors had about the relationship between their personal and professional sustainability-related motivations, behaviours and actions.

Personal Sustainability Actions of Participants

As shown in Table 19.1, all participants engaged in a variety of personal and professional sustainability activities. In terms of personal actions, common to all was recycling, minimising food waste, sustainable forms of travel, minimising purchasing and taking the stairs rather than the lift. It was apparent that participants A and B, who identified themselves as "lead users", were engaged in personal sustainability actions to a greater extent than participants C and D (the "regular users"). For the lead users, the consideration of sustainability issues and impacts were part of their daily lives and affected most, if not all, of the decisions they made. That is not to suggest that participants C and D were less aware of sustainability; on the contrary, all participants perceived themselves as knowledgeable on sustainability matters and recognised its importance. However, participants C and D recognised there was a greater disconnect between their knowledge and actions: as participant C remarked, "I'd say it's super important but my actions probably mean it's more like moderately important or somewhat important if I was being very, very honest." All participants acknowledged the difficulty of converting their knowledge and importance that they attached to sustainability into action, and a number of challenges (i.e. unsustainable behaviours) were identified such as eating meat and fish, purchasing international food (e.g. exotic fruits), owning (non-electric) cars, and a penchant for travelling (flying), and shopping (new clothes).

Professional Sustainability Actions of Participants

As with personal sustainability behaviours, participants A and B engaged with sustainability in their professional role to a much greater extent than did participants C and D. Whilst all participants implemented individual sustainable behaviours as seen in their personal lives (e.g. recycling, avoiding waste, taking the stairs), the lead users engaged with, and promoted, sustainability at multiple levels and with different stakeholders in their professional environment, too. Not only had both A and B developed and led new modules on sustainability,

Table 19.1   Participants' personal and professional sustainability actions

	Personal	Professional
Participant A (female, 35–44, UK, 4-year tenure, lead user)	• Recycles • Minimises purchases • Avoids unethical companies • Pescatarian • Energy efficiency • Minimises car journeys • Ethical banking • Minimises flight travel (carbon offset-ting when does)	• Recycles • Avoids printing • Takes stairs • Walks to work • Teaches sustainability / ethics modules • Developed two sustainability modules • Participates in sustainability forums • Advises faculty on sustainability in teaching • Researches sustainability
Participant B (female, 45–54, Portugal–Germany, 3.5-year tenure, lead user)	• Recycles • Minimises purchases • Donates clothes and buys second-hand clothes • Engages in "LOFE" purchasing behaviour: local, organic, free-trade, environmentally sustainable • Walks / uses public transport (no car) • Carbon offsetting for flights • Campaigns for sustainability issues	• Recycles • Incorporates sustainability extensively in teaching • Developed sustainability-related modules and staff training • Encourages faculty to incorporate sustainability in teaching • Participates in sustainability forums • Advises students societies and teams on sustainability • Researches sustainability
Participant C (male, 25–34, UK, 4-year tenure, non-lead user)	• Recycles • Zero food waste • Walk/cycles • Minimal purchases	• Recycles • Avoids printing • Walks/cycles to work • Takes the stairs (not lift) • Limits travel to international conferences • Instigated sustainability research project at DMU
Participant D (female, 25–34, Thailand, 4.5-year tenure, non-lead user)	• Recycles • Donates to causes (e.g. animal welfare, nature conservation) • Uses public transport (no car) • Highlights inequalities with family/friends (e.g. racial/gender inequality)	• Recycles • Takes the stairs (not lift) • Commutes via public transport • Highlights sustainability in teaching • Instigated sustainability research project at DMU

but they were involved in promoting sustainability at the faculty and organisational level. For example, both were active members of the university-wide Education for Sustainable Development Forum (a cross-faculty forum for academic staff with an interest in embedding sustainability), and had advised faculty on how to implement sustainability in teaching. However, of the two lead users, participant B was more involved in communicating and driving sustainability within the university and faculty. For instance, she contributed to a university-wide project exploring the integration of sustainability and the SDGs in the curriculum and had promoted sustainability amongst students outside of her formal teaching duties. In addition to this, participant B was the only one (of the four participants) who focused explicitly on the UN SDGs and related topics in her teaching and research, finding them "really easy for me to structure and explain a number of sustainability issues to others." Whilst A and B (the lead users) were more professionally involved, it is interesting to note that in late 2019, participants C and D (the regular users) initiated a study focusing on sustainability at DMU (e.g. drivers, implementation, key stake-holders), and then invited participants A and B to collaborate on the study. The rationale for this was initially one of opportunity, given the topicality of sustainability within DMU strategies. Accordingly, sustainability represented an effective way to explore a theory of interest to both participants, and as staff members this provided relatively easy access to empirical data. C and D expressed that they had since developed an interest in sustainability and are keen to see how the university's commitment materialises and what role they will play. These changing sustainability attitudes and behaviours are in line with the lead user construct, which is not dichotomous (thus not defining if an individual is either a lead user or not) but can be gradual. Additionally, the construct is not an inherent trait but a characteristic that develops and disappears over time. Accordingly, a lead user in a given context "may or may not be a lead user in that domain in the future" (Hienerth & Lettl, 2017, p. 4).

Reflections on the Incorporation of Personal Experiences within Faculty Role

Owing to the autoethnography approach adopted for this chapter, the authors engaged in some very frank exchanges about the challenges and issues asso-ciated with incorporating personal (sustainability) values and actions into our professional work. We believe it is important to highlight and discuss some of these debates as it is crucial that we question and reflect on our practices and taken-for-granted assumptions in relation to communicating and promoting sustainability.

The first issue was the impact and effectiveness of academic staff highlight-ing their personal actions and experiences in their professional role. In relation to students, participant A commented that she uses her sustainability expe-riences extensively in her teaching and worried that she shared "maybe too much" of herself. Participant A further explained that she focuses particularly on communicating the challenges she faces with trying to be more sustainable (e.g. owning a "fossil fuel guzzling car") as she believes it is important to show that sustainability is a complex issue, and that being open about this on a personal level will encourage students to discuss their own challenges. This also raised a further concern of participant A; namely, that she may come across as "preaching" about sustainability, which, she believes, can produce the opposite effect in students and colleagues. Whilst participants C and D (the regular users) did not explicitly use their sustainability experiences in their interactions at work (for participant C, this was because he was "not an authority for that topic"), they did use examples from their private life in their teaching (where relevant) as they believed it helped to establish a relationship with students. Additionally, participants B and D believed that sharing their personal experiences (on a range of issues) was particularly beneficial for students, in that, owing to their multi-cultural backgrounds, this encouraged the sharing of different views and ways of seeing the world. Indeed, we all agreed that creating this "diversity in viewpoints" (as participant B stated) is important for issues and problems that are complex, contested, and multi-faceted, such as sustainability and the SDGs.

The second revolved about whether and how staff should incorporate their personal experiences into their academic roles to influence staff. Participant A, for example, commented that she had not used her role as programme leader / course director to promote sustainability and the SDGs amongst staff that teach on her programme, for fear of "preaching" and/or being seen to push her own agenda and interests. Participant B, on the other hand, had discussed with staff on her programme about sustainability and the SDGs but with a 'light touch'; namely, asking them to consider embedding sustainability and the SDGs into their teaching along with other initiatives. The reason for taking this more moderate and encouraging approach came from a concern that academic staff were under a considerable amount of workload pressure and may resist additional demands. Participant D highlighted that staff who draw on and shared their personal interests and experiences (be it sustainability or some other issue) would not come across as 'preachy' or put staff under pressure if this was a genuine interest and passion of the person sharing this. Linked to this, she further noted that "we do need people who are advocates that are also genuine because it brings courage out of other people who might be on the fence."

DISCUSSION

This study, and its unique approach (that being, four academics reflecting on their engagement with sustainability), has resulted in three key insights. First, the study found a disconnect between the personal and professional behaviours of the study's participants. Whilst all four authors engaged in sustainability actions in their personal lives, for two of the participants (A and B), these actions had become a significant part of their professional role. In relation to why participants C and D had not engaged more extensively with faculty sustainability activities, it is important to highlight that this was not connected to a principle of sharing personal experiences at work. Indeed, all authors believed this was beneficial, especially for enhancing teaching practice (such as engaging students on an issue/topic). Rather, they felt they lacked the experience as well as personal interest in sustainability (compared with A and B), and that their 'voice' would be perceived by others as lacking credibility and authenticity.

The importance of authenticity represented this study's second key insight for how personal sustainability actions and beliefs can impact sustainability in the workplace. Being authentic – which relates to having a deep understanding of, and being true to, one's values and beliefs – was considered a crucial 'trait' or characteristic for those who are, or aspire to be, lead users for sustainability within their organisations (Haddock-Fraser et al., 2018). However, for authenticity to be fully effective, it was felt that change agents must also be courageous in the sense of being true to their (personal) values and displaying these beliefs at work. Academics that show courage in their (sustainability) convictions, it is argued, can help empower and inspire others to adopt similar behaviours and/or find the courage to promote and influence sustainability in their sphere of influence.

The third and final key insight that emerged from this study relates to the different workplace behaviours of the two lead users. Although both were very knowledgeable and attached great importance to sustainability, one of the lead users was more involved in contributing to the university and business school's sustainability activities. It is interesting to note that, despite these differences, they both *perceived* themselves to be lead users and 'ahead of the curve' in relation to embedding sustainability in the workplace. This highlights that perceptions differ about what the lead user role entails, and may also suggest that an archetype lead user does not exist and, indeed, may not be desirable. For universities to fully embrace sustainability, it might require different types of lead users, ranging from academics who focus on particular aspects of their work (e.g. teaching/curriculum) to those who concentrate on broader university mechanisms and processes (e.g. contributing to university-wide committees and projects to shape the agenda of the institution). Despite the variation in the lead users' engagement with sustainability, this study suggests that personal authenticity and courage are aspects that all types of lead users require in order to effectively drive sustainability in the workplace.

CONCLUSION AND FUTURE RESEARCH

This chapter's findings suggest that the lead user concept is a useful construct to explain the mindset, behaviours and actions of faculty staff who are key drivers in changing the paradigms taught, researched and enacted within universities, in the pursuit of a sustainable future. This study extends the lead user concept by, first, applying it to a public sector context, and viewing individual employees as potential lead users. The suitability of the lead user construct to HE is underlined by its explicit acknowledgement of the potential changing nature of lead user characteristics. Second, by purposefully involving those who do not self-identify as sustainability advocates or lead users, the resulting insights are more nuanced and serve as a more realistic starting point into further phases of planning and action.

Whilst extending the literature, this chapter has some limitations. First, its generalisability is limited by the small number of participants. Its autoethnographic approach, although unveiling interesting nuances, carries the risk of biased conclusions. More in-depth work on the positionality of the individual participants could be valuable. Future studies could extend beyond 'lead user characteristics' to consider the 'lead user method', and analyse the short- and long-term impact of educators' sustainability actions on multiple stakeholders as well as various contexts. Overall, given today's wicked problems and grand challenges, and the scope of the SDGs and their potential multi-dimensional impact, this research area will continue to merit attention.

REFERENCES

Alvesson, M., Hardy, C., & Harley, B. (2008). Reflecting on reflexivity: Reflexive textual practices in organization and management theory. *Journal of Management Studies*, *45*(3), 480–501.

Ansari, S., Fiss, P. C., & Zajac, E. J. (2010). Made to fit: How practices vary as they diffuse. *Academy of Management Review*, *35*(1), 67–92.

Ansari, S., Reinecke, J., & Spaan, A. (2014). How are practices made to vary? Managing practice adaptation in a multinational corporation. *Organization Studies*, *35*(9), 1313–1341.

Argyris, C., & Schön, D. A. (1978). *Organizational Learning: A Theory of Action Perspective*. Boston, MA: Addison-Wesley.

Barth, M. (2013). Many roads lead to sustainability: A process-oriented analysis of change in higher education. *International Journal of Sustainability in Higher Education*, *14*(2), 160–175.

Barth, M., & Rieckmann, M. (2012). Academic staff development as a catalyst for curriculum change towards education for sustainable development: An output perspective. *Journal of Cleaner Production*, *26*(2012), 28–36.

Bilgram, V., Brem, A., & Voigt, K.I. (2008). User-centric innovations in new product development – Systematic identification of lead users harnessing interactive and collaborative online-tools. *International Journal of Innovation Management, 12*(3), 419–458.

Ellis, C., Adams, T. E., & Bochner, A. P. (2011). Autoethnography: An overview. *Historical Social Research/Historische Sozialforschung, 36*(4), 273–290.

Franke, N., von Hippel, E., & Schreier, M. (2006). Finding commercially attractive user innovations: A test of lead-user theory. *Journal of Product Innovation Management, 23*(4), 301–315.

Haddock-Fraser, J., Rands, P., & Scoffham, S. (2018). *Leadership for Sustainability in Higher Education*. London: Bloomsbury Publishing.

Hart, S., & Ahuja, G. (1996). Does it pay to be green? An empirical examination of the relationship between emission reduction and firm performance. *Business Strategy and the Environment, 5*, 30–37.

Hienerth, C., & Lettl, C. (2017). Perspective: Understanding the nature and measurement of the lead user construct. *Journal of Product Innovation Management, 34*(1), 3–12.

Kanashiro, P., Rands, G., & Starik, M. (2020). Walking the sustainability talk: If not us, who? If not now, when? *Journal of Management Education, 44*(6), 822–851.

Lettl, C., Hienerth, C., & Gemuenden, H. G. (2008). Exploring how lead users develop radical innovation: Opportunity recognition and exploitation in the field of medical equipment technology. *IEEE Transactions on Engineering Management, 55*(2), 219–233.

Lilien, G. L., Morrison, P. D., Searls, K., Sonnack, M., & von Hippel, E. (2002). Performance assessment of the lead user idea-generation process for new product development. *Management Science, 48*(8), 1042–1059.

Menon, S., & Suresh, M. (2020). Synergizing education, research, campus operations, and community engagements towards sustainability in higher education: A literature review. *International Journal of Sustainability in Higher Education, 21*(5), 1015–1051.

Meyer, J. W., & Rowan, B. (1977). Institutionalized organizations: Formal structure as myth and ceremony. *American Journal of Sociology, 83*(2), 340–363.

Nonaka, I., & Konno, N. (1998). The concept of "Ba." *California Management Review, 40*(3), 40–54.

OECD. (2015). *Oslo Manual: The Measurement of Scientific and Technological Activities*. https://www.oecd.org/science/inno/2367614.pdf (accessed 31 May 2021).

Rock, P. (2007). Symbolic interactionism and ethnography. In P. Atkinson, A. Coffey, S. Delamont, J. Lofland, & L. Lofland (Eds.), *Handbook of Ethnography* (pp. 26–38). London: Sage.

Schumacher, M. C., & Kuester, S. (2012). Identification of lead user characteristics driving the quality of service innovation ideas. *Creativity and Innovation Management, 21*(4), 427–442.

Times Higher Education. (2019). *Impact Rankings 2019*. https://www.timeshighereducation.com/rankings/impact/2019 (accessed 2 October 2020).

von Hippel, E. (1986). Lead users: A source of novel product concepts. *Management Science, 32*(7), 791–806.

von Hippel, E. (2005). *Democratizing Innovation*. Cambridge, MA: MIT Press.

Wijen, F. (2014). Means versus ends in opaque institutional fields: Trading off compliance and achievement in sustainability standard adoption. *Academy of Management Review, 39*(3), 302–323.

Wu, Y. J., & Shen, J.-P. (2016). Higher education for sustainable development: A systematic review. *International Journal of Sustainability in Higher Education, 17*(5), 633–651.

20. OS4Future: an academic advocacy movement for our future

Giuseppe Delmestri, Helen Etchanchu, Joel Bothello, Stefanie Habersang, Gabriela Gutierrez Huerter O, and Elke Schüßler

INTRODUCTION

OS4Future is an international movement of organization and management scholars that seeks to act on the climate emergency.[1] Ours is a movement that complements a larger international group of scholars called Scientists for Future, who support the youth movement Fridays4Future as inspired by the Swedish activist Greta Thunberg. OS4Future, founded in 2019, seeks to create and inspire change through a focus on personal and academic activities: research, teaching, practice, and leading by example. Our modus operandi is to "walk the talk" when it comes to sustainability.

As organizational scientists, we are in a unique position to offer solutions to tackle societal challenges. We understand the opportunities associated with organizational and institutional transformation, as well as the complexities and challenges at a cognitive and behavioral level. We also benefit from being embedded within business schools; in this capacity, we can help reshape the purpose of business, improve decision-making processes, develop policies, and raise awareness among political and business leaders and society at large.

Yet at the core of the OS4Future movement is a focus on personal actions that we, as scholars, take towards sustainability. These actions are targeted at SDG 13 (the Sustainable Development Goal on Climate Action) related to strengthening resilience and adaptive capacity to climate-related hazards, integrating climate change measures into national policies, and improving education, awareness-raising, and human and institutional capacity on climate change mitigation and adaptation. These actions have symbolic power, where we model the behavioral changes that prefigure a zero-carbon world.

THE ORIGINS

Our movement was initiated through the personal observation of a problem in our profession: the majority of our colleagues traveled to conferences by plane, oftentimes selecting the most time- and/or cost-efficient itinerary, which unfortunately ended up being the most carbon intensive. However, we were also inspired by solutions, namely one member's "sabbicycle" (sabbatical on a bicycle). We proceeded to consider constructive solutions to reduce the carbon footprint of conference travel; towards this end, in the spring of 2019, we launched the initiative #EGOSbytrain for that year's EGOS (European Group of Organizational Studies) conference. We sought to provide a more environmentally friendly alternative to plane travel: 33 scholars signed a pledge to attend EGOS by train, posting on social media about the initiative. This communication signaled the support of colleagues who were also seeking to reduce their carbon footprint.

Four of us traveled together by train from London to Edinburgh, which gave us the time and space to engage in a prolific brainstorming session about future actions. We were four scholars from four different countries on a 4-hour train ride. This precious collective time allowed us to launch the OS4Future movement – we created the mission to be a "movement of organization and management scientists who wish to inspire fellow academics to take action on climate change … focus[ing] on actions that make a concrete impact on four dimensions: research, teaching, practice, leading by example" (OS4Future, n.d.). The OS4Future movement thereby specifically targeted SDG 13 of climate action, to "take urgent action to combat climate change and its impacts" (United Nations, n.d.).

As of November 2020, OS4Future has a network of 54 affiliated individual endorsers comprising international junior and senior academics from a range of business schools and universities.[2] While concentrated in Europe, we also have members from many other regions and have a steadily growing digital presence (at the time of writing, 842 and 53 academic followers on Twitter and Facebook respectively, in addition to our individual networks). In the following section, we summarize our actions on research, teaching, practice, and leading by example at different levels (from individual to societal) and discuss various challenges of walking the sustainability talk.

ACTIONS

"Walking the Talk" in Our Personal Lives

Even before the movement, we were all self-reflective to varying degrees about reducing our environmental impact through concrete changes: for instance, many of us opted to live car-free, choosing to cycle or use public transport instead. In line with the Flying Less pledge,[3] we also reduced flying and chose train travel wherever possible. At home, we adopted a range of diverse actions, whether consuming a plant-based diet, reducing plastic waste, buying local produce, growing fruit and vegetables, making eco-efficiency improvements at home, and minimizing overall material consumption. These have provided a repertoire of successful actions (as well as challenges) that we can share with others who are interested in similar changes.

Yet, we recognize that we are subject to social and physical constraints. For instance, some of us relocated abroad and are distant from families that live on other continents. Others live in rented apartments and cannot make home improvements; some raise young kids, which severely restricts time for other activities; while others live in remote areas with limited access to public transport. We are not, however, complacent in these constraints; rather we see this movement as a journey of continuously examining our personal choices and making changes wherever we deem it possible.

Prefiguration in Our Home Institutions

Members of OS4Future have also been active in driving new pedagogical initiatives at their home institutions. A first activity has been transforming the curriculum: one of us has designed a "Sustainability Management" major/minor at her school, while others have developed new modules including a "Business Ethics & Sustainability" course at the undergraduate level and an MBA level course called "The Responsible Manager". We have also reshaped the content of existing modules to reflect the urgency of tackling climate change by, for instance, including SDG-inspired content. We have also participated in Lectures4Future (an initiative of Scientists4Future in Austria, Germany, Italy, and Switzerland aimed at teaching each other's students in a new course format), performed Lectures4XR,[4] and shared educational resources with colleagues.[5]

Some of us have assumed administrative roles at our schools and voluntarily joined committees and working groups across the university; this has included coordinating a Sustainability Lab or participating in a university-level procurement and sustainability working group. Through these efforts, we can champion and push sustainability policies, e.g. selecting catering services that are vegan and locally sourced, or increasing the offer of plant-based meals at canteens.[6] Other key developments in this area have been the creation of policies and pledges for academic traveling such as promoting a policy of ground travel for destinations within 12 hours of home, or a university-level Flying Less pledge.[7]

What is evident is that some universities are more advanced and open to implement these policies while others are more conservative. For some of us the first and most critical step has been simply to be outspoken to senior management about the importance of these policies and to spark debate. To communicate best practices in teaching and university policies, we have created open documents on our website compiling these ideas. Finally, senior members of OS4F have been advocating to value service and professional impact work for tenure and promotion decisions.

Sustainability Engagement in Our Professional Communities and Beyond

We have all been cognizant about choosing research areas that are consistent with our mission, whether that involves topics (e.g. climate social movements, climate policy, inequality, sustainability and markets, climate change and coffee value chains, sustainability education in business schools, business ethics in the relationship to animals) or methods (e.g. action research, experimental methods, participant observation). Members of OS4F have also started to publish together (Etchanchu, de Bakker, & Delmestri, 2021). Most of us are part of or in the process of creating research chairs on the sustainability transition.

In tandem, we have taken up leadership positions in professional associations and are currently involved in the organization of relevant special issues to further the debate and contribute to a research agenda (e.g. "Organizing Sustainably" in *Organization Studies*; "Sustainable Digital Innovation" in *Creativity and Innovation Management*). Moving beyond the boundaries of our professional communities, we also engage with civil society and policy-makers through participatory research (for instance with Extinction Rebellion, a social movement advocating against climate change and biodiversity loss) and one of our colleagues has applied to join the Canadian Federal Government's Sustainable Development Advisory Council.

To inspire others about our message, we communicate our actions and their associated tensions in different ways. For example, the OS4Future website publishes Stories4TheFuture,[8] which feature faculty anecdotes of climate action on different dimensions. Some of us add the label "member/co-founder of OS4Future" to our academic signatures, which lends visibility to the movement. We have experienced that this signaling provides more legitimacy to speak about the SDGs with the general public, and has led to interviews and national media presence. For example, one of us participated in the French national debates on the occasion of the five-year anniversary of the SDGs, which brought together various stakeholders, including academics, businesses, civil society actors, and politicians. We also talk about our own individual commitments in the classroom to encourage students and illustrate concretely how they can make a difference. We find that our personal sustainability actions lend us credibility and add to the students' learning experience by inspiring emulation. We also actively engage with them in campus greening activities (e.g. using reusable cups and bottles in cafes, reducing printing, advocating for less carbon-intensive travel).

CHALLENGES AND RESPONSES

Gathering Support from the Academic Community

Colleagues from within and outside our institutions have, to varying degrees, pushed back against our initiative, or expressed doubts. As an example, after one of us mentioned that he had traveled by train, an academic peer responded: "Good, more space on the plane for me." These reactions were not entirely unexpected, as other scholars have documented similar experiences of behavioral inertia (e.g. Hoffman, 2016; Kanashiro et al., 2020). Some colleagues have discounted the efforts of the movement altogether, arguing that piecemeal actions like ours have marginal effects on solving the larger issue; others simply avoid taking part in a potentially uncomfortable spotlight on their unsustainable academic and personal lifestyles. Through personal experience, we have found that confronting people is rarely effective. Skeptics and those who have entrenched behaviors become defensive if they feel they are being judged.

In light of this challenge, OS4Future has pursued a multi-pronged "pincer" approach, using different methods and platforms to engage in collaborative conversations with our colleagues (Klonek, Güntner, Lehmann-Willenbrock, & Kauffeld, 2015). For one, to raise awareness and garner new endorsements for our initiative, OS4Future members integrate their efforts in existing association events. We have been speakers in panels of various Professional Development Workshops (PDWs), and PhD and Early Career Workshops at the 2019 Academy of Management (AOM), and have also organized a sub-plenary at EGOS 2020. For EGOS 2021, we follow up with another sub-plenary entitled "OS4Future: Walking the Talk and Talking the Walk about the Climate Crisis", as well as with a sub-theme: "Generativity through engaged scholarship: Connecting theory, methods & praxis". We believe that these spaces provide arenas to debate and to raise deeper questions about our role as academics in contributing to (un)sustainability and the gradual normalization of the environmental crisis (Blühdorn, 2013). In order to specifically address some of the challenges for junior scholars (see below), one of us was a keynote speaker during the 2020 EGOS PhD and Early Career Workshop in the topic of "Sustainable Academic Lives" and, as co-organizer, supported a shift in the Early Career Workshop away from discussing research papers towards discussing "paths to a meaningful career".

In parallel, we organize standalone events. Prior to the London–Edinburgh train journey in 2019, we held a research seminar and round table discussion organized at the institution of one of the co-founders. The seminar was planned in an interactive way and involved a paper presentation by one of the founders on his work on normativity, action research, and institutional theory, which then would serve as framing for a discussion on the symbolic power of behavioral changes, prefiguring a zero-carbon world, and on the role of organizational scientists. The event was attended by roughly two dozen enthusiastic colleagues from surrounding institutions.

We have also discovered that combining efforts with allies in other faculties is effective – one of us participates in a university-wide sustainability center, including a Climate Emergency Committee headed by the Geography Department. Our strategy is to find internal allies in other departments who may be more receptive of the "flying less" agenda and securing participation in strategic committees in which sustainability policies are pushed more aggressively. Another colleague is applying a pincer movement to his efforts to influence the university to adopt more sustainable practices: he is silently supporting a group of Students4Future (providing local information, publication resources, editing texts) and openly addressing the same issues in the University's sustainability center and meetings of the academic work council.

Leveraging Sustainable Individual Commitments towards Impactful Change

While changing academic and personal lifestyles choices will help to reduce our footprint, we have also started to mobilize with other collectives in academia. These include the Group for Research on Organizations and the Natural Environment (GRONEN), Academy of Management Organizations, and the Natural Environment division (AOM ONE), Scientists4Future, the Impact scholar community. One of our co-founders is a guest convenor within the newly created EGOS Standing Working Group "Organization Science in the Anthropocene: System Change, Not Climate Change", which will run through 2021–2024.

The collective dynamic of the OS4Future movement has inspired and fuels individual-level commitments and initiatives. For example, we learned that The Impact Scholar Community drew inspiration from OS4Future. The positive dynamic of bringing like-minded colleagues together sparks positive social contagion, which encourages substantive actions in our respective individual contexts.

Navigating Individual and Various Professional Commitments at Different Career Stages

Although the path of engaged scholarship is meaningful, we are admittedly often overwhelmed by multiple commitments and role conflicts. Unfortunately, given the incentive system in which we operate, with constant pressure to publish in top journals, we are regularly trapped in trade-offs; we feel that climate action may hinder our careers since it is not sufficiently valued by our professional community. One member has written about how this system reinforces the "impostor syndrome", a phenomenon that undermines self-confidence in one's abilities (Bothello & Roulet, 2019). Clearly, our research time is diminished if we start to engage more broadly in climate action. One junior member openly commented: "I wonder whether I'm not screwing up my career." But the same problems hold for scholars at all career stages, usually faced with multiple research, teaching, community, and university service demands – on top of private ones – against which they need to weigh off their climate action. These tensions were already experienced by sustainability management scholars in the early 1990s, when the ONE division at the AOM was just emerging (Stead & Stead, 2010).

Another of our colleagues, a junior member of her university, engages in a lot of introspection and reflection on how she wants to spend her time:

> I didn't choose an academic career to be wrought up with external pressures that nudge me into activities that I don't really want to spend my time on; it feels like an enormous waste of valuable time and intellectual resources. Often, I wonder: what if all those smart people around me thought about how to address the most pressing issues of our times around the SDGs rather than spend years on framing and reframing a research study to make a theoretical contribution with a paper whose empirical insights could have been published years ago .... What would we say about medical research on the coronavirus that isn't published for the next 5 years because the authors are still tweaking the theoretical framing of the study? I see a lot of junior colleagues around me who get into academia because they want to change things and share this deep sense of impact.

These reflections "vividly show the systemic nature of the problems we must confront". As Kanashiro and colleagues (2020, p. 838) comment, junior faculty "should be aware that ... delaying research on important but difficult topics until one has tenure, can result in the establishment of habits and life patterns that are very difficult to change once one finally has the freedom to do so." Junior colleagues require psychological and emotional reassurance that, despite the obstacles they face, they are pursuing the right path. Clearly, we need this to be a collective effort and platforms such as the impact scholar community or our OS4Future movement may hopefully help legitimate individual sustainability actions.

One of us who recently completed her PhD also experienced the dilemma of keeping her individual environmental footprint low versus following "the rules of the game." In academia, participation in prestigious international conferences, e.g. the Academy of Management (AOM) is crucial for building a network and receiving feedback. In addition, participation also has an important signaling effect, serving as a powerful heuristic for hiring institutions or funding organizations to evaluate the "quality" of an early career scholar's work when their research is not (yet) published. However, with regular attendance of 10–12,000 scholars, the AOM conference is a significant source of CO_2 emissions. For Europe-based scholars, attendance necessitates long-distance flights: a return flight to the US from Germany emits more than three tons of CO_2, or 150% of the climate compatible annual budget per person.[9] Hence, the decision to attend international conferences or not is a real dilemma for young scholars:

> If I want to stay in academia do I need to attend such a conference at least once or twice? Can I ever be part of the solution (through educating students about sustainable business practices) when I was part of the problem (by engaging in flying to conferences)? Or should I avoid conferences such as AOM completely and take the risk that a hiring committee or a funding institution might conclude that I wasn't "good enough" to get accepted there?

Clearly, there is not an easy answer here and trade-offs must be carefully evaluated on an individual level. For example, another Europe-based member visited the AOM conference twice at the beginning of her PhD, but from then on only participated in conferences and workshops that could be reached by public ground transport. She favors EGOS as a European conference and regularly checks if leading scholars participate in summer schools, workshops, or lectures in Europe. Such smaller local events turn out to be much more productive, allowing for easier contact with peers and senior scholars and higher quality of received feedback.

Nevertheless, going to conferences by public transport also comes with challenges, especially for early career scholars. First, it is more time-consuming and may involve some negotiation with supervisors for longer absences. Second, using public transport is sometimes more costly with many universities covering only the cheapest travel option (which is often a plane ticket). While the former issue might be resolved more easily – as a long train ride can be quite productive for writing – the latter issue may be harder to resolve. Some PhD students might afford additional costs for train travel by themselves, but many others are less privileged.

What might be a good argument to help PhD students secure increased travel funding for ground-based travel? Before the COVID-19 pandemic forced EGOS organizers to offer the 2020 conference online, the OS4F team had planned to organize pre-conference workshops on trains leading to Hamburg. The idea was to use the voyage as an event in which scholars could discuss their research and work together on specific topics. For example, the initiative #EGOSbyTrain was a semi-official event that was also supported by the conference organizers on their webpage. Inspired by this initiative, colleagues who had already endorsed OS4Future organized a common train trip to a New Institutional Conference in Denmark just before the pandemic forced lockdowns and confinement. Such events may help many PhD students to better justify the ground-based travel options. Of course, more frequent online conferences may facilitate conference attendance thanks to decreased funding requirements.

We conclude this section by also pointing to the pushbacks that senior scholars may encounter for their visible sustainability actions in business schools. One of us was confronted by colleagues with questions like, "Do you always need to study these topics?" or comments like, "If the administration reduces our travel budget because of your climate-activism it will be your fault." We recognize that avoiding flights reduces opportunities for additional income (like executive education or consultancy abroad) that are relatively easy to obtain for a high-status business school professor and are part of the normal expectations and financial planning at this later career stage. These preoccupations are not comparable with the anxieties of junior scholars but may explain the rationalization of a high-carbon lifestyle and the difficulty for some to walk the talk. Certainly, senior academics can also gain a lot by living in a way that is aligned with one's own values, as well as by concentrating on other priorities such as community and family.

Strategic Redirection Following the COVID-19 Pandemic

With the COVID-19 pandemic, we observe how the business of "high flying business schools" (Gill, 2020) marked by intense international student mobility and "academics on an aeroplane" (Parker & Weik, 2014) is grinding to a sudden halt, making virtual modes of conferencing and teaching a necessity, rather than an inferior second-best option. In response, the main focus of our activity has shifted from advocating for and organizing sustainable conference traveling towards trialing out virtual and blended formats for academic scholarly exchange and teaching. The crisis is being seized as an opportunity by university policymakers to reconsider their practices for distance learning and, as more universities come to understand international mobility of both faculty and students as a main driver for university-related carbon emissions, develop new guidelines. An open dialogue about promising practices will be urgently needed in this process, since we neither want universities to close themselves off to international exchange nor to return "back to normal". We recognize that many faculty members and students alike suffer from "Zoom fatigue" and realize how important in-presence classroom debates and informal interactions – among and within student and faculty communities – are for a productive work climate. Yet, many also come to value the "affordances" provided by digital tools, such as being able to participate in workshops or conferences without having to travel or reserving valued classroom time for discussions based on previously shared online lectures or materials. These are new opportunities for those who were previously excluded from such events – a realization that global academia is itself stratified (Murphy & Zhu, 2012) along potentially harmful status dimensions and that, if challenges are global, it is necessary to tackle problems differently at the regional and global level. Simultaneously, given that we and our supporters are largely based in Europe and North America, the movement is striving to reach other academic communities in the Global South whose experience of the climate emergency and the pandemic may be substantively different from ours.

We need fresh ideas for blended conference and teaching formats and to share this knowledge across universities, e.g. the detailed report on the virtual "Organization Studies Summer Workshop" one of our members co-organized (Schüßler, Helfen, Pekarek, Zietsma, & Delbridge, 2020), or the blog post by Howard Aldrich titled "What I learned about virtual conferences from AOM 2020" (Aldrich, 2020). We can also use digital technology to develop new forms of collaboration in teaching, such as the open access course "Organizing in times of crisis",[10] developed by one of our members together with other colleagues from different universities to offer scholars around the world more up-to-date, research-driven course materials for teaching during the COVID-19 crisis. While our basic mission has not changed since the COVID-19 crisis, our prefigurative organizing practices have extended further into conference organizing in a more sustainable way towards gathering best practices in online teaching.

CONCLUSION

We have presented the origins, actions, and quandaries of the young academic movement of OS4Future. The increasing attention and sympathies that this movement is generating are pushing us in a potentially virtuous circle; we feel tremendously indebted to all our supporters and hope to generate adherence of like-minded scholars to fuel collective action. Despite the described challenges to combine our mission with the demands of academic and university life, we are growing increasingly confident that OS4Future may help legitimate and fuel individual and collective sustainability actions to have a broader impact towards achieving the SDGs.

NOTES

1. At https://os4future.org/ (accessed May 31, 2021; unless otherwise stated, all websites were visited on that date).
2. To endorse OS4Future, please visit https://os4future.org/contact/
3. See https://academicflyingblog.wordpress.com/
4. See Scientists for Extinction Rebellion, https://www.scientistsforxr.earth
5. For example, SDGAcademy.org or https://utopiaplatform.wordpress.com
6. See https://os4future.org/policy-recommendations-for-conference-travelling/
7. See https://os4future.org/universities-business-school-travel-policy-examples/
8. See https://os4future.org/stories-4-the-future/
9. At https://www.atmosfair.de/en/
10. At https://timesofcrisis.org

REFERENCES

Aldrich, H. (2020). What I learned about virtual conferences from AOM 2020, at https://howardaldrich.org/2020/08/what-i-learned-about-virtual-conferences-from-aom-2020/ (accessed May 31, 2021).

Blühdorn, I. (2013). The governance of unsustainability: Ecology and democracy after the post-democratic turn. *Environmental Politics*, *22*(1), 16–36.

Bothello, J., & Roulet, T. J. (2019). The imposter syndrome, or the mis-representation of self in academic life. *Journal of Management Studies*, *56*(4), 854–861.

Etchanchu, H., de Bakker, F. G. A., & Delmestri, G. (2021). Social movement organizations' agency for sustainable organizing, in Teerikangas, S., Onkila, T., Koistinen, K., & Mäkelä, M. (eds), *Research Handbook of Sustainability Agency*. Cheltenham, UK and Northampton, MA, USA: Edward Elgar Publishing, forthcoming.

Gill, M. (2020). High flying business schools: Working together to address the impact of management education and research on climate change. *Journal of Management Studies*, *58*(2), 554–561.

Hoffman, A. J. (2016). Reflections: Academia's emerging crisis of relevance and the consequent role of the engaged scholar. *Journal of Change Management*, *16*(2), 77–96.

Kanashiro, P., Rands, G., & Starik, M. (2020). Walking the sustainability talk: If not us, who? If not now, when? *Journal of Management Education*, *44*(6), 822–851.

Klonek, F. E., Güntner, A. V., Lehmann-Willenbrock, N., & Kauffeld, S. (2015). Using Motivational Interviewing to reduce threats in conversations about environmental behavior. *Frontiers in Psychology*, *6*, 1015, at https://www.ncbi.nlm.nih.gov/pmc/articles/PMC4508486/ (accessed May 31, 2021).

Murphy, J., & Zhu, J. (2012). Neo-colonialism in the academy? Anglo-American domination in management journals. *Organization, 19*(6), 915–927.

OS4Future (n.d.). At https://os4future.org/ (accessed May 31, 2021).

Parker, M., & Weik, E. (2014). Free spirits? The academic on the aeroplane. *Management Learning, 45*(2), 167–181.

Schüßler, E., Helfen, M., Pekarek, A., Zietsma, C., & Delbridge, R. (2020). Reimagining conferencing – Organizing the #OSSW20 sustainably. Utopia Platform for Imagining Transformations, at https://utopiaplatform.wordpress.com/2020/06/06/reimagining-conferencing-organizing-the-ossw20-sustainably/ (accessed May 31, 2021).

Stead, J., & Stead, W. (2010). Sustainability comes to management education and research: A story of coevolution. *Academy of Management Learning & Education, 9*(3), 488–498.

United Nations. (n.d.). Goal 13, at https://sdgs.un.org/goals/goal13 (accessed May 31, 2021).

21. The tie that binds: how economic literacy is a foundation for sustainability
Madhavi Venkatesan

ACADEMIC SUSTAINABILITY

Globally, economies are measured in relative terms with respect to a single economic indicator, the gross domestic product (GDP). This indicator, which was created to measure output, has become a synonym for the standard of living, in direct opposition to the caution related to the same put forward by its creator, Simon Kuznets (1934). Further, quantification of economics in the decades that followed its deployment has successfully facilitated the perception of economics as a science and distanced the discipline from its moral philosophical roots (Chandavarkar, 2007; Dow, 2010). Instead of the behavioral discipline of economics facilitating policy based on an explicit normative framework of seeking to attain societally optimal outcomes related to ecosystem symbiosis and quality-of-life attainment, economics in practice has been molded by mathematics to equate optimality with maximum production capacity (Gowdy and Erickson, 2005). The latter, however, is an implicit normative judgement that has promoted business as usual as solely a fiduciary responsibility. Indeed, in present practice economics characterizes failures of the market mechanism to optimize social outcomes as "externalities", while the teaching of economics has endogenized marketing and profit-making by legitimizing that individuals are insatiable and driven by greed. That these assumptions, which are referenced as theory, are social norms borne into existence through social construction are not addressed (Beach, 1938; Earle et al., 2017).

With 70 years of this standardized teaching, the economy has been molded by theory and its behavior is consistent with insatiability, high resource use, excess production and consumption, as well as a lack of acknowledgment for non-market costs, such as the responsibility for waste. Arguably, this reality provides an opportunity for the discipline to acknowledge its relevance with respect to issues affecting global environmental health. It is the production-driven model along with an exclusion of non-market responsibility that has created the social and environmental justice issues of the present, including poverty and climate change (Venkatesan and Keidl, 2021).

Even though a behavioral science, economics has been taught as though it is an optimization discipline, and that incentives like utility and profit maximization are immutable drivers of human behavior. Students are typically presented with a single perspective that reinforces the economic framework in operation. However, the teaching also bounds the rationality of the student to the extent that there is no critical assessment of the prevailing economic system, no discussion of moral responsibility of fundamental purpose other than accumulation. As early as 1938 Beach noted,

> It has been the policy of many teachers of economics that beginners in economic theory should be taught only one theoretical explanation of each phenomenon. These teachers feel that the acquaintance of the student with other explanations can be postponed until the student has become more familiar with economics in general and therefore has a better sense of judgment …. Perhaps the most important job of the teacher in social sciences is to develop the students' power of discernment. The students must learn that one idea does not contain the whole truth; and when this is learned, the students' progress will be more rapid. (p. 515)

In the present period, researchers have highlighted that the teaching of economics has not deviated much over the past 50 years, stating that the profession of the teaching of economics has lagged with respect to alignment of content with contemporary issues (Allgood et al., 2015; Bowles and Carlin, 2020). This is a lost opportunity and a problem. The exclusion of application and limited tangibility in the classroom teaching of economics has limited recognition of the significance of economic literacy. Given that introductory economics is typically a requirement at most institutions, non-tangible and commoditized teaching does not allow the profession the opportunity to be relevant.

More broadly, a few courses in undergraduate economics, and perhaps only an introductory course, are often the only interaction that the college graduates of tomorrow will have with the economics profession. As they are the only opportunities that academic economists will have to educate the citizens and voters of tomorrow, they deserve our best efforts (Becker, 2000, p. 117).

ECONOMICS OF SUSTAINABILITY

In 2017, I developed an interdisciplinary course at Northeastern University entitled *Economics of Sustainability*. The course outline is provided in Table 21.1 and reflects my interest to make economics both tangible and connected to the issues related to sustainability. The opportunity in the teaching of

Table 21.1   *Economics of Sustainability course outline*

Module 1	Understanding the Economic Framework
Module 2	Limitations of GDP
Module 3	Externalities: Life Cycle Assessment Production, Consumption and Waste
Module 4	Sustainability: Holistic Evaluation
Module 5	Corporate Social Responsibility, Investor Activism and Grassroots Movements
Module 6	Paris Agreement and the Sustainable Development Goals: Economics of Climate Change
Module 7	Sustainability and Economic Progress: Reframing Purpose

economics is based on the role of economics in transforming our planet. This discussion references the adoption of a production-based metric (GDP) and the separation of non-market supply chain impacts (i.e. exploitation of labor, environmental degradation) from daily transactions. The course is divided into seven modules as provided in Table 21.1.

In each module there is a direct connection between economics and sustainability, and wherever possible I use documentary film and outside speakers to increase student understanding, visualization, and connection between the topics and life on the planet. The overall goal of the curriculum is to facilitate the course learning objectives, which are stated in Table 21.2.

Although Module 6 addresses all the Sustainable Development Goals (SDGs), the discussion is related to an understanding of each individual SDG and the alignment of the SDGs with areas of national concern, with the latter being assessed in relation to the Paris Agreement. Module 3 directly addresses SDG 12: Responsible Consumption and Production and promotes an understanding of the interrelationship between regulation, production and consumption. What I highlight is the common element in all three entities: the individual economic agent. Individually in a democratic system, we all have an opportunity to vote,[1] to determine where we invest our money and to decide how we consume. I argue that these attributes have power, respectively, to affect government leadership and regulation, supply chain management and sustainable production, and the brand premium attributed to sustainable production.

To supplement this discussion, students engage in an individual life cycle assessment where they evaluate the life cycle impact of their favorite beverage, inclusive of the container. They assess the qualitative resource footprint of both the beverage and container in separate assessments from extraction/ planting, to consumption, and disposal, and compare overall cycle impacts to the price paid for the product. In the final stage of the assessment, they provide a justification for the current price or suggest an alternative price based on the qualitative impacts they uncover and the risk tolerance they assume for infor-

Table 21.2   Economics of Sustainability course learning outcomes

1.	Students will be able to state the relationship between culture and economic outcomes and will understand the significance of intergenerational social construction as it relates to existing economic systems and thought.
2.	Students will understand the evolution of economic thought and the influence of economic assumptions on consumption and production behavior.
3.	Students will understand the significance of the attribution of "rational" to economic agent behavior and the importance of economic literacy in the attainment of rational behavior.
4.	Students will be able to identify externalities and identify the mechanics of market failure as they relate to sustainability.
5.	Students will be able to define sustainability in relation to economic systems. Students will also be able to discuss the concept of sustainability from the perspective of conscious consumption and will be able to appreciate how values fundamentally determine economic outcomes.

mation yet unknown. Typically, this assignment affects student understanding and potentially even behavior but most importantly makes tangible the impact of transparency and information symmetry in determining the true cost of their consumption good. Effectively the assignment also introduces students to the significance of responsible consumption and production (SDG 12) in establishing sustainability.

In spite of my passion for sustainability integration and my interest in connecting my students to the relationship between their values and behaviors, and observable economic outcomes, *Economics of Sustainability* is a challenging course to teach. The topic of the course mimics the dynamics of the global economic framework and solutions require deliberate action. However, students expect to be provided with established tools and remedies and are not satisfied with the role of individual and collective action in reconfiguring the focus of economic growth to be explicitly tied to well-being (Cohen, 2017). Instead, they find it difficult to imagine an alternative economic framework, perhaps in part due to a lack of exposure in curriculum and through cultural experience.[2] As a result, I find that there is a need to provide examples of both activism and tangible solutions to ensure that students remain engaged. This is where my non-profit activities offer value and overlap with my teaching. My non-profit activism and leadership provide a present reminder of Margaret Mead's quote: "Never doubt that a small group of thoughtful, committed, citizens can change the world. Indeed, it is the only thing that ever has." I provide a background on my non-profit, our activities, and the change we have championed in the next section.

Table 21.3   Sustainable Practices, mission statement

The overarching mission of Sustainable Practices is to facilitate a culture of sustainability as defined by reducing the anthropogenic impact to the planet and its ecosystems.

At the time of its incorporation and designation as a federally recognized non-profit, Sustainable Practices had one objective, to increase consumer awareness of the relationship between consumption and environmental and social justice issues. Presently, we are engaged in three ongoing community initiatives:

- The Cape Plastic Bottle Ban campaign, a grassroots effort to promote understanding and civic engagement with respect to the adverse consequences of single-use plastic bottle consumption on both human and environmental health. The Ban initiative was announced in March 2018 and has been covered by multiple news outlets.
- Sustainable Practices Sustainability Film Series is a documentary screening series started in the fall of 2017. Due to COVID-19 our films are presently screened online and are available on-demand.
- Sustainable Practices, a podcast and radio show, is hosted by WOMR and airs the first Friday of every month at 9:00 am. Podcasts are available for listening on-demand under Events on the WOMR website.

CIVIC SUSTAINABILITY

I established Sustainable Practices in 2016 in Barnstable County (Cape Cod), Massachusetts. The organization's mission is to promote sustainability by reducing the human impact on the planet; see Table 21.3. Cape Cod is made up of 15 towns (Barnstable, Brewster, Bourne, Chatham, Dennis, Eastham, Falmouth, Harwich, Mashpee, Orleans, Provincetown, Sandwich, Truro, Wellfleet and Yarmouth) and is an island accessible by bridge and ferry from mainland Massachusetts. Known for its beaches and temperate climate, Cape Cod is a tourist destination and tourism is a significant source of seasonal revenue. Primarily visitors to Cape Cod are from Massachusetts and the neighboring states of Connecticut, New York and New Jersey. The island is becoming a second-residence community, with increasingly more homeowners having a primary residence elsewhere.

As part of our routine civic engagement we maintain four Adopt-A-Highway sites and one Adopt-A-Visibility site on highway 6 on Cape Cod. This activity along with our seasonal All-Cape Beach Clean-up give me the opportunity to see the litter problem and specifically the adverse consequences of plastic pollution on the Cape first hand.

Although I introduce my students to all the activities of Sustainable Practices, the Cape Plastic Bottle Ban, given that it is a presently ongoing action, provides them with real-time understanding of citizen action and how commercial activity can be deliberately changed to promote sustainability. The in-class discussion of the Bottle Ban efforts provides my students with a stronger understanding of the challenges of promoting change including, the educational and communication outreach required, and the confrontations from individuals who may disagree. They are able to see how Upton Sinclair's (1935) comment, "It is difficult to get a man to understand something, when his salary depends on his not understanding it", is in fact a significant limitation to needed change for sustainability.

Cape Plastic Bottle Ban

The Cape Plastic Bottle Ban effort started in 2018 with the documentary screening of *Divide in Concord*, a film that traced the three-year history of the first commercial single-use plastic water bottle ban in the United States in Concord, Massachusetts. In 2019, we initiated the Municipal Plastic Bottle Ban, which focused on eliminating non-emergency single-use plastic bottle purchases by town governments and sale of beverages in single-use plastic containers on town property across all 15 towns in Barnstable County, Cape Cod. As of year-end 2019, the Municipal Plastic Bottle Ban was adopted by 11 of the 15 towns that comprise Barnstable County. In 2020, we facilitated the filing of the Municipal Plastic Bottle Ban in Bourne, Mashpee, Truro and Barnstable and initiated the Commercial Single-use Plastic Water Bottle Ban (Commercial Ban) in the 11 towns that adopted the Municipal Plastic Bottle Ban. The Commercial Ban eliminates the sale of non-carbonated, non-flavored water in single-use plastic bottles of less than 1 gallon in size within the jurisdictional area of a town. As of year-end 2020, in spite of COVID-19, which postponed town meetings in several towns, thirteen towns had adopted the Municipal Ban and seven towns had adopted the Commercial Ban.

Given that in the state of Massachusetts all government meetings are taped and accessible to the public through town websites, I am able to share the dynamics of the entire process: filing a bylaw; meeting with stakeholders, including town leadership; and the final debate and vote on the town meeting floor. Students are able to engage with the effort and are able to see how change can be made.

Connecting Academic and Civic Engagement

What is perhaps the most synergist aspect of my civic engagement with respect to my academic work is the discussion related to the life cycle impact of single-use plastic bottle consumption (see Figure 21.1) given that this is also an element of the *Economics of Sustainability* course.

The Bottle Ban effort is promoted through educational outreach that includes local speaking events with and hosted by community organizations, churches and schools among others, as well as requisite attendance at meetings with town government leaders. As a routine part of these meetings, I address the life cycle impact of single-use plastic bottle consumption highlighting

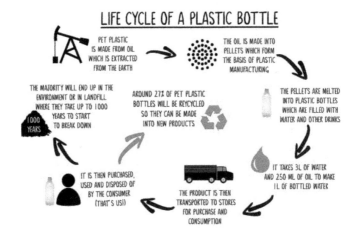

Figure 21.1   Life cycle of a plastic bottle

the qualitative impacts at each major segment of the product life as noted in Table 21.4. I limit discussion of quantitative impacts given the challenge of empirical replication of research findings, typically due to the small sample size and contextual differences of studies. A debate on the values related to pollution damage allows the discussion of the adverse impacts to deviate from the intention.

However, even after addressing the life cycle impacts other issues remain. Active push-back centers on the following themes:

- Bottled water is safer than tap.
- Bans are eliminating my freedom.
- Merchants will lose a significant source of revenue.
- People are going to drink unhealthy beverages.
- State or federal ban implementation is more effective.

Each of these stated items offers an additional teaching opportunity aligned to the interdisciplinary nature of *Economics of Sustainability*, while the life cycle provides a tangible example of the application of SDG 12. Further, responses to push-back themes are aligned to a direct or indirect aspect of economics. The assumptions related to bottled water highlight the power of marketing, while the use of bans for behavioral change relates to the use of economic limitations by the government to promote the greater good (i.e. taxes on cigarettes, illegalization of cocaine). The loss of merchant revenue highlights the opportunity to discuss how revenue presently generated is subsidized by present and future human and environmental health; and the assumption of

Table 21.4   *Life cycle assessment of single-use plastic*

Production:	What is the impact of plastic production on the climate?
Consumption:	What are the impacts of single-use plastic on human health?
Disposal:	Isn't litter the problem and can't recycling help?

substitution between beverages is based on a faulty perception that water can be substituted with another beverage. Economics assesses substitution affects and is able to use cross-price elasticity to determine the potential for substitution for water. Finally, the perception of a top-down view of regulatory change offers the opportunity to highlight the limitation of this policy through the lens of historical time and also highlight the significance of grassroots movements in the establishment of regulation that has been successful. Information related to all these themes is readily accessible both in the academic literature and increasingly in the credible press.

INTEGRATING THE SUSTAINABLE DEVELOPMENT GOALS

My non-profit work and my classroom instruction of sustainability are mutually reinforcing. My efforts on Cape Cod provide tangible, real-time evidence of how individuals working together can make a difference. This discussion allows me to address global issues from a local impact perspective but also allows me to transition to discussions of local issues across the globe and the fundamental handicap to sustainability, poverty. Given that poverty is the focus of the SDGs, I can address a few issues and action items with my classes that are also applicable to a wider global constituency. Further, I can apply the same discussion points to my *Economics of Race* and *Economics of Crime* classes, where poverty is observed as a racialized social norm.

Perhaps the most significant observation I provide my students, though, is that poverty cannot be defined by income alone. To the extent that income can only address the market or transaction value of human existence it omits the qualitative attributes such as the emotional benefits of having an ability to interact in society, meet with friends, and have a consistent home. In this manner SDG 1: No Poverty is related to SDG 8: Decent Work and Economic Growth, as decent work, in a market economy, would likely need to be in excess of subsistence to allow an individual to partake in society (Venkatesan and Luongo, 2019). Similarly, SDG 12: Responsible Production and Consumption also includes a relationship with poverty that affects the degree of social and environmental justice issues. Responsible production considers the holistic impacts rather than only regulatory restrictions and in so doing does not take advantage of disenfranchised and marginalized communities. Responsible consumption seeks to eliminate information asymmetry and include supply chain activities in purchasing decisions. In this way the consumer chooses to purchase goods that are not subsidized by adverse environmental and social impacts to vulnerable populations. This discussion parallels the life cycle assessment incorporated in the Bottle Ban efforts as it provides me an opportunity to highlight the impact and responsibility inherent in production and consumption if we are to attain sustainability.

In highlighting the relationship between the present economic model and the sustainability issues of focus in the SDGs, I am able to address the role of values. To the extent that we view immediate gratification and assume fair market prices, our purchase decisions align with production that can be solely focused on meeting regulatory limitations rather than the totality of externalities. If we value environmental and social justice along with economic equity, we have a responsibility to eliminate supply chain externalities, including exploitation of people and the planet. Finally, I am able to question the efficacy of using the present economic system to achieve sustainability: If we use the same methods to take us out of our current problem that were used to get us into it, are we being realistic? Do we need a change in our values or a combining of our values into our decision-making, with this being a social norm? These are questions with answers that quiet contemplation and active discussion can provide. They also set the next stage of the discussion, the value-based paradigm shift to enable sustainability as an individual behavior and a social norm.

NOTES

1. The issue of mass incarceration in the United States limits the enfranchisement of racially marginalized communities who in turn are disproportionately represented in poverty. Poverty continues to be a discriminatory factor that affects choice limits participation of individuals (Alexander, 2020; Edelman, 2017).
2. In course evaluations students have noted a desire for solutions related to environmental degradation but are not satisfied with the perspective of modifying the prevailing economic structure to align with sustainability. Comments have noted that values specific to consumption and accumulation are not possible to change.

REFERENCES

Alexander, M. (2020). *New Jim Crow: Mass Incarceration in the Age of Colorblindness.* New York: New Press.
Allgood, S., Walstad, W., & Siegfried, J. (2015). Research on teaching economics to undergraduates. *Journal of Economic Literature*, 53(2), 285–325.
Beach, E. (1938). Teaching economics. *The American Economic Review*, 28(3), 515.
Becker, W. E. (2000). Teaching economics in the 21st century. *The Journal of Economic Perspectives*, 14(1), 109–119.

Bowles, S., & Carlin, W. (2020). What students learn in economics 101: Time for a change. *Journal of Economic Literature*, 58(1), 176–214.

Chandavarkar, A. (2007). Economics and philosophy: Interface and agenda. *Economic and Political Weekly*, 42(3), 221–230.

Cohen, S. (2017). Democratic grassroots politics and clean economic growth. *Journal of International Affairs*, 71(1), 103–116.

Dow, S. (2010). Enlightening economics. *RSA Journal*, 156(5544), 24–27.

Earle, J., Moran, C., Ward-Perkins, Z., & Haldane, A. (2017). Beyond neoclassical economics. In *The Econocracy: The Perils of Leaving Economics to the Experts* (pp. 60–91). Manchester: Manchester University Press.

Edelman, P. B. (2017). Not a crime to be poor: The criminalization of poverty in America. *The Guardian*, 6 November.

Gowdy, J., & Erickson, J. (2005). The approach of ecological economics. *Cambridge Journal of Economics*, 29(2), 207–222.

Kuznets, Simon S. (1934). National Income 1929–1932. Senate Document No. 124, 73rd Congress, 2nd Session. Washington, DC.

Sinclair, U. (1935). *I, Candidate for Governor: and How I Got Licked.* Los Angeles, CA: End Poverty League.

Venkatesan, M., & Keidl, N. (2021). Addressing the relationship between economics and climate change: A discussion of principles. *Sustainability and Climate Change*, 14(1), 35–41.

Venkatesan, M., & Luongo, G. (2019). *SDG8 – Sustainable Economic Growth and Decent Work for All.* Bingley, UK: Emerald Publishing.

Index